NEURAL
NETWORK
DATA
ANALYSIS
USING
SIMULNET™

T0213921

Springer

New York
Berlin
Heidelberg
Barcelona
Budapest
Hong Kong
London
Milan
Paris
Santa Clara
Singapore
Tokyo

Edward J. Rzempoluck

Neural Network Data Analysis Using Simulnet™

With 45 Figures

Springer

Edward J. Rzempoluck
Brain-Behaviour Laboratory
School of Applied Sciences
Simon Fraser University
Burnaby, BC V5A 1S6
Canada

Library of Congress Cataloging-in-Publication Data
Rzempoluck, Edward J.
    Neural network data analysis using Simulnet / Edward J. Rzempoluck.
        p.   cm.
    Includes index.
    Additional material to this book can be downloaded from http://extras.springer.com.

    ISBN 0-387-98255-8 (alk. paper).
    1. Neural networks (Computer science).   2. Numerical analysis—Data
processing.   3. Simulnet.   I. Title.
QA76.87.R94   1997
519.5′01′13632—dc21   97-16666

Printed on acid-free paper.

Production managed by Terry Kornak; manufacturing supervised by Johanna Tschebull.
Photocomposed copy prepared by Springer, using the author's Microsoft Word files.

9 8 7 6 5 4 3 2 1

ISBN 0-387-98255-8   Springer-Verlag New York Berlin Heidelberg   SPIN 10628729

Acknowledgments

My sincere thanks to the members of the Brain Behaviour Laboratory at Simon Fraser University for their many helpful comments and criticisms during the development of Simulnet. I want to thank in particular Dr. Harold Weinberg, director of the Brain Behaviour Laboratory, for his guidance and for his emphasis on the importance of defining the substantive question at the beginning of a research project.

Parts of this book had their origins in a set of laboratory exercises for a biological psychology course at Simon Fraser University. To the students who labored through those exercises, my thanks for their patience and the useful feedback they provided.

I want to thank Dr. Martin Gilchrist at Springer for taking a chance on this book, and Springer's production department for its patience with my many revisions to the text and to the software.

Finally, I want to express my gratitude to my darling wife for her insight, help and patience with the editing of the book.

Contents

Introduction.. 1
    Scope of this Text 1
    What Is Expected from the Reader 3
    An Outline... 3
    Computer Requirements... 4

1   The Simulnet Desktop **5**
    Introduction... 5
    Desktop Components 5

2   Data Analysis... **13**
    Introduction... 13
    The Substantive Question 15
    Neural Network Analysis 15
    Genetic Algorithms and Neural Networks 75
    The Probabilistic Network 91
    The Vector Quantizer Network.. 102
    Assessing the Significance of Network Results 113
    Network Application Examples 116
    Fractal Dimension Analysis 128
    Fourier Analysis 144
    Eigenvalue Analysis.. 146
    Coherence and Phase Analysis.. 153
    Mutual Information Analysis 160
    Correlation and Covariance Analysis.. 163

3   Acquiring and Conditioning Network Data **171**
    Introduction... 171
    Data Specification 171
    Data Collection.. 173
    Data Inspection 177
    Data Conditioning 180
    Detrend—Order 0 184
    Standardize Columns 185
    Frequency Filtering 187
    Principal Component Analysis.. 189
    Principal Component Data Reduction... 199

4   A Data Analysis Protocol.. **203**

Introduction... 203
A Preprocessing Checklist 203
Analyzing Experimental Data.. 204

Glossary **213**

Index.. **219**

Introduction

Scope of this Text

This text is intended to provide the reader with an introduction to the analysis of numerical data using neural networks. Neural networks as data analytic tools allow data to be analyzed in order to discover and model the functional relationships among the recorded variables. Such data may be empirical. It may originate in an experiment in which the values of one or more dependent variables are recorded as one or more independent variables are manipulated. Alternatively, the data may be observational rather than empirical in nature, representing historical records of the behavior of some set of variables. An example would be the values of a number of financial commodities, such as stocks or bonds. Finally, the data may originate in a computational model of some physical process. Instead of recording variables of the physical process, the computer model could be run to generate an artificial analog of the physical data.

Since data in virtually any native form can be expressed in numerical format, the scope of the analytical techniques and procedures that will be presented in this text is essentially unlimited. Sources of data include research work in a range of disciplines as diverse as neuroscience, biomedicine, geophysics, psychology, sociology, archeology, economics, and astrophysics.

An often fruitful approach to data analysis involves the use of neural network functions. Neural network functions discussed in this text include multilayer feedforward networks trained using the error backpropagation algorithm; neural network-genetic algorithm hybrids; generalized regression neural networks, learning vector quantizer networks; and self-organizing feature maps.

The obvious question is then, how can such neural networks be of use in the analysis of data? Most significantly, neural networks can behave as universal function approximators. As such, neural networks are capable of modeling the functional associations among the independent and dependent variables in the data. As an example of this capability, neural networks can be used as pattern classifiers. They can be trained to assign a pattern of values of a set of independent variables, to one of a number of categories.

The application of a number of other data analytical techniques will also be discussed. These techniques have been selected for inclusion on the basis of the following criteria. First, the techniques can be usefully applied as adjuncts to neural network analysis, performing any one of a variety of preprocessing operations on the data. Second, the techniques are some that are not generally discussed at a level that is accessible to a wide range of nonspecialist users.

Techniques in the first category, those intended to perform a data preprocessing function for subsequent network analysis, include correlation and covariance analysis, eigenvalue analysis, and Fourier analysis. Correlation and covariance are both estimates of the strength of linear association between a pair of variables. Eigenvalue analysis provides an estimate of the complexity inherent in the data by finding a set of features, mutually independent combinations of the measured variables, that can account for a majority of the variance in the data. Fourier analysis, like eigenvalue analysis, also repackages the variance in the data. In the case of Fourier analysis however, the repackaging is done in terms of predefined features; sine and cosine functions. Essentially, Fourier analysis acts as a mathematical prism. As a prism separates a beam of white light into its component colors, so Fourier analysis shows the magnitudes of the different frequency components within the data.

Techniques in the second category include fractal dimension analysis, mutual information analysis, and coherence analysis. Fractal dimension analysis can, under suitable conditions, provide an estimate of the complexity of the multivariate dynamical system from which the data has been sampled. Coherence and mutual information are both measures of association. Mutual information is an index of the degree of general, rather than simply linear, association between two data sets. If the data represent a time series— the data values are recorded at some interval over a span of time—then mutual information can be viewed as an index of general association in the time domain. Coherence is an analogous measure in the frequency domain. Coherence is an index of the degree of association between the respective frequency components in two data sets.

All of these analytic techniques can be used in order to directly address questions about the data. In addition many of these procedures can be used to preprocess data that will then be analyzed using a neural network. Data preprocessing includes operations carried out on data that are intended to reduce the computational or analytic load on subsequent analyses, in particular, analyses using neural networks. Preprocessing can simultaneously achieve a number of different goals.

First, preprocessing the data can reduce the computational load on a network. Such a reduction may not only be convenient, but it may be critically important if the analysis has to be done within a limited time span. Neural networks are inherently parallel processes. As such they can easily become effectively unusable when implemented on a serial machine and, if at the same time, they are asked to deal with a large volume of data.

Second, preprocessing can reduce the analytic load on a network. When presented with suitably preprocessed data, the network only needs to deal with a subset of the features within the data. These features would be ones that are hypothesized to be relevant to the analytic task at hand. For example, if the task is to classify electroencephalographic (EEG) recordings, the experimenter might entertain the hypothesis that the experimental manipulation has had an effect on the EEG in terms of between-channel correlations. It might be useful to preprocess the data by computing between-channel correlations. The volume of data represented by these correlations would be small in comparison with the volume of the raw EEG data. The neural network would then be trained on only the correlations, rather than on the raw data.

In sum, data preprocessing can help the neural network by selecting from the entire mass of data those components that the researcher has independent reason to believe may be most relevant to the analytic question. To the extent that the selected features are pertinent to this question, the neural network will often have a much easier job of learning to model the data.

Supporting the discussions of the various data analytic techniques, a number of examples and case studies will be presented and discussed. The procedures in the examples can be carried out using the software supplied with this text, Simulnet. Simulnet is a program that is designed to carry out a wide range of analytical, transformational, and graphical operations on numerical data. Simulnet shares some features with traditional statistical software, but extends the capabilities of such software by offering access to analytically powerful neural network-based functions. The significance of such neural network-based functions is that they allow data to be explored for relationships or patterns that might not be discoverable using more traditional statistical procedures.

What Is Expected from the Reader

The reader is expected to bring to this text two skills. The first skill involves computer literacy. The reader is expected to be familiar with the operation of Windows-based software. Readers without this background can prepare themselves by working through one of the many texts that are available to introduce novice users to the Windows operating system. The second skill involves mathematical background. This text generally assumes a mathematical background on the part of the reader that is equivalent to a first-year course in calculus and related topics.

An Outline

- Section 1 presents a description and explanation of the components of the Simulnet desktop.
- Section 2 discusses approaches to data analysis, with respect to the types of questions that are correspondingly addressed. This section will describe a range of analytical techniques, and will focus on the application of neural networks to the analysis of data. Each of these analytical approaches is illustrated using a number of preprepared examples that demonstrate in detail how these functions are used.
- Section 3 addresses the issue of how readers can use the analytical functions in Simulnet to explore their own data. This discussion will involve the data visualization and transformation functions that Simulnet provides in order to allow data analysts to preprocess and condition their data prior to analysis.
- Section 4 presents a data analysis protocol; a generalized procedure that can be followed in a first attempt at examining a set of data.
- Section 5 is a glossary of some of the more complex terms used in this text.

Computer Requirements

In order to work through the exercises in this text using Simulnet, the following hardware and software configuration is assumed:

- Software: Windows 3.1 or newer.
- Hardware: minimum, 386 CPU with 8 Mbytes of RAM; highly recommended, 486DX or later CPU with 16 or more Mbytes of RAM.

Information about upgrades and new releases of Simulnet can be obtained at the following web address:

http://www.springer-ny.com/supplements/simulnet

1
The Simulnet Desktop

Introduction

Simulnet is a program that is designed to carry out mathematical operations on numerical data. In contrast with traditional spreadsheet or statistical software packages, Simulnet provides analytical power that is not generally available with such software. These functions are based on a set of neural network and genetic algorithm-based algorithms. Simulnet operations can be grouped into four general categories: Data transformation, analysis, visualization, and modeling. This introduction includes exercises that allow the reader to explore a sample of the functions that are available in each of these four categories.

Starting Simulnet

To install Simulnet, run the file *Setup.exe* located on disk 1 or on the CD-ROM. All data files that are used in the examples in this text will be located in the Data subdirectory of the directory in which Simulnet was installed. Path names in this text will assume that Simulnet was installed to the default directory *Simulnet*. If Simulnet is installed to a different path, substitute the name of that path whenever a path name is mentioned in the text.

Context-sensitive on-line help is available at all times. Help can be invoked at any time by pressing the *F1* key on the keyboard, or by clicking on the *Help* button on a dialog form. The help facility contains detailed instructions on the use of all Simulnet functions.

Desktop Components

When Simulnet has finished loading, the desktop should look as shown in Figure 1.1, but without the matrix or graph forms shown. To reproduce Figure 1.1 carry out the following steps:

1. From the *File* menu, select the *Open File* option to show the *Open File* dialog form. From the *\Simulnet\Data* directory, select file *test1.dat*. This will open the file as a matrix. The matrix should appear on the desktop as an icon. If the matrix form appears in normal size, minimize the form to an icon.

2. From the *Graph* menu, select the *XY Graph* option. On the *XY Graph* dialog form, click the *New Graph* button. A graph form should appear on the desktop. This graph form should resemble that shown in the figure. On the *XY Graph* dialog form, click the *Close* button to remove the dialog form from the desktop.

3. The desktop should now contain only the matrix icon and the graph form. Restore the matrix form to normal size. The desktop should now resemble Figure 1.1.

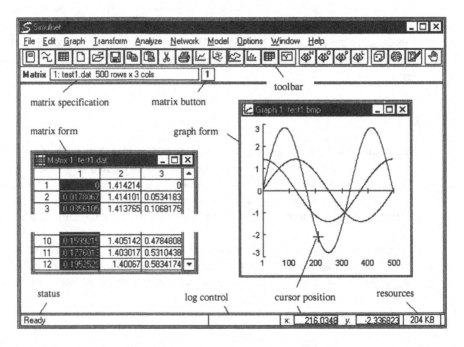

Figure 1.1. This view shows the locations of the desktop components. The desktop is shown containing a typical matrix form and a graph of the data in the matrix columns.

Table 1.1 describes the various components of the Simulnet desktop.

Table 1.1 Simulnet Desktop Components

Component	Description
Matrix Specification	The matrix specification displays information about the matrix on the desktop that is currently active. If no matrix is currently active, this display is blank. This situation may occur if there are no matrices on the desktop or if there are matrices on the desktop, but none of them currently is active. To choose a matrix on the desktop as the active matrix, click the left mouse button anywhere on the matrix form or icon, or click the appropriate matrix button.
	The matrix specification consists of the matrix number and the number of rows and columns for the matrix. The matrix number is a reference number that corresponds to the number in the title bar on the matrix form. Each new matrix form is assigned a unique matrix number as it is created on the desktop. For example, if the currently active matrix on the desktop is a matrix of 20 rows by 30 columns, with the title "Matrix 3: [Untitled]", the matrix specification display will show the following message:
	Matrix 3: Untitled 20 rows x 30 cols
Matrix Button	A matrix button is created for each matrix form that is created on the desktop. The number on the button is the matrix number of the associated matrix. When the desktop contains more than one matrix form, clicking a matrix button will make the matrix in the associated matrix form the focus of subsequent operations.
Toolbar Buttons	The toolbar buttons provide shortcut access to a number of Simulnet functions.
Status Message	The status message appears on the left-end of the status bar. This message indicates the current activity. If no activity is currently taking place, the displayed message is "Ready." Certain Simulnet functions will show their progress with a percentage figure, indicating the state of completion of the function.
Resources Display	The resources display indicates both the amount of free memory in the system, and the percentage of free system resources. Only one of these values is shown at any time. To display the alternate value, click the display once. Clicking the display once more will return the display to the original state. This display normally indicates amount of free memory.

The amount of free memory is an indication of the size of matrix that can be created. A matrix requires approximately (4 x rows x columns) bytes of memory. Divide this number by 1,024 to find the memory requirement in kilobytes (KB).

The percentage of free system resources indicates the capacity remaining in Windows to create screen objects (for example, matrices, text-boxes, graphs, etc.)

Log control

The log control allows session logging to be paused and resumed, and indicates the current state of session logging. Note that this control is disabled when session logging is not in progress.

Cursor Position
Display

When an xy graph is present on the desktop, the cursor position display appears on the status bar. When the mouse cursor is moved across the drawing surface of the graph (the graph base), the shape of the mouse cursor changes to a cross-hairs pattern. The cursor position display indicates the position of the center of the cross-hairs on the graph drawing surface. Cursor position is shown in the true coordinates of the graph. The resolution of this indication is limited by the number of pixels on the graph since the mouse cursor can only move in increments of one screen pixel. A larger graph, with more pixels in the x and y directions, can therefore be examined with more resolution than the same graph plotted on a smaller graph.

Toolbar Buttons

Toolbar buttons provide shortcut access to Simulnet functions.

Button	Description
	displays a dialog form allowing a log of the current Simulnet session to be started. The log is a disk file containing a continuously updated record of operations carried out in Simulnet, and the results of those operations. A copy of the log is maintained in a text-box on the desktop entitled the Session Log. The contents of the Session Log can be examined at any time to see the results of past operations.
⎍	displays the *function generator* dialog form. The function generator can be used to create artificial data, containing sine, cosine, and rectangular functions. Optionally, pseudo-random noise can be added with either a uniform or Gaussian distribution.

displays the *Matrix Order* dialog form that allows the specification of the number of rows and columns for a new, blank matrix.

creates a new, blank text box.

displays the *Open File* dialog form that allows a disk file to be opened as a matrix, textbox or bitmap.

saves a matrix, text, or graph to disk. If the item has not been previously saved, or if the item currently exists on disk, a dialog form will be presented providing path and file name options.

Copies the selected portions of the matrix to a Simulnet Buffer, and optionally to the Windows Clipboard, when the selected desktop object is a matrix form. The Windows Clipboard has a capacity of approximately 60,000 characters. The capacity of the Simulnet Buffer is limited only by the amount of memory available to Windows. When the amount of data to be copied exceeds the capacity of the Clipboard, data can be copied to the Simulnet Buffer. Data copied to the Clipboard is available to other Windows programs. Data copied to the Simulnet Buffer is available only within Simulnet. When the selected object is a text form, the selected portion of the text is copied to the Clipboard. When the selected object is a graph, the graph is copied to the Clipboard.

allows numerical or text data to be pasted into a matrix or text form. The numerical data may originate from a matrix form or in a text form. Note that pasting text data into a matrix form may result in values of zero being pasted in: Text evaluates to zero.

cuts the selected data from a matrix or text form to the Windows Clipboard.

prints the selected matrix, text, or graph using the printer and settings defined using the *Printer Setup* option on the *File* menu.

displays the *XY Graph* dialog form.

displays the *XYZ Graph* dialog form.

displays the *Surface-Graph* dialog form.

displays the *Multi-Graph* dialog form.

displays the *BackProp Network* dialog form.

displays the *Genetic Network* dialog form.

 displays the *Probabilistic Network* dialog form.

 displays the *Vector Quantizer Network* dialog form, providing access to both the supervised (Learning Vector Quantizer [LVQ]) and unsupervised (self-organized map [SOM]) training modes.

 randomizes network weights (Neural, Genetic, and Vector Quantizer networks).

 initiates network training or resumes paused training.

 initiates network testing.

 pauses network training. Clicking this button will pause training of the BackProp Network, Genetic Network, and the Vector Quantizer Network. Training can then be restarted at a later time with no loss of information (assuming Simulnet has not been shut down). This button also terminates a number of other analytic functions. When such functions are terminated, the results of the analysis are generally not available.

Desktop Options

A number of options are provided to allow desktop functions to be customized. One of these options should be configured at this time. From the *Options* menu, click the *Show new matrix as icon* option. This will select this option, as indicated by a check mark beside the option. Selecting this option will cause all new matrices that are brought onto the desktop to appear initially as icons. These matrices can be restored to the normal window state at any time. By showing all new matrices as icons when they are first created, the desktop is kept from becoming too congested. If the *Show new matrix as icon* option is not selected (no check mark beside it) all new matrices that are created on the desktop will appear in the normal ('restored' in Windows terms) form. See on-line help for more information on this options.

Matrices: Units of Data in Simulnet

Data in Simulnet is handled in the form of matrices. A matrix is simply a table of numbers; that is, a collection of numbers organized into rows and columns. In common with usual data-analytic practice, Simulnet assumes that matrix columns contain variables. In other words, one matrix column is assumed to contain the data for a single variable. Correspondingly, rows are assumed to contain the cases or data-points associated with each of the column variables.

As an example, one column of a matrix might contain the values of a set of measurements of body temperature for one subject. This column thus contains the values of the variable *body temperature*. There might be more than one such set of measurements of body temperature. In that case, each set of measurements would be stored in its own column. Suppose that temperature measurements are available for each of five subjects. For each subject there six temperature measurements, one measurement taken on each of six days. These values could then be stored in a matrix of six rows (one row for each measurement) by five columns (one column for each subject). This matrix will be denoted by **X** (matrices are generally represented by bold-face upper-case letters). Matrix **X**, containing hypothetical values in degrees Celsius, would appear as follows:

Row (temp)	Column (subject) 1	2	3	4	5
1	40.1	35	41.3	41.9	38.4
2	38.8	40.5	40.7	37.6	37.1
3	39.2	40.9	37.7	38.8	39.5
4	37	40.1	42	40.6	39.7
5	37.1	35.2	41.3	35.3	36.8
6	40.6	38	35.3	39.3	37

$$\mathbf{X} =$$

Figure 1.2. Matrix X contains temperature measurements for each of five subjects Data for each subject are contained within one column. Each row contains a single temperature measurement. This matrix has been provided in file *intro1.dat*.

Simulnet Exercise: Opening a File as a Matrix

As an exercise in manipulating matrix forms on the desktop a file will be opened as a matrix, the matrix will be inspected, and the matrix will then be removed from the desktop. Note that whenever this text indicates that a file should be opened, the file is meant to be opened as a matrix unless it is specifically stated otherwise. To open a file as a matrix, select the *Open File* option from the *File* menu. Alternatively, click on the *Open File* toolbar button. This will show the *Open File* dialog form, allowing the file to be opened to be selected.

1.   From the *File* menu, select the *Open File* option. On the *Open File* dialog form select the directory in which the Simulnet data files were installed. By default, this will be *\Simulnet\Data*. Simulnet data files will always be initially installed in a subdirectory of the main Simulnet directory, named *\Data*. From this directory select file *intro1.dat*.

2. Click the *OK* button. A matrix icon should appear on the desktop. The matrix specification display should show the following information: *Matrix 1: intro1.dat 6 rows x 5 cols*. Matrix 1 has 6 rows and 5 columns. This is the first matrix form created on an empty desktop, so it has been given the matrix number of 1. Note that a corresponding matrix button has appeared immediately to the right of the matrix specification, labeled to correspond to the matrix number 1.

3. Restore the matrix form to normal size. The matrix should be displayed in the matrix form as a table of rows and columns, with table entries as shown in Figure 1.2.

    Since this matrix has a small number of rows and column, the matrix is shown in the matrix form in its entirety. If the matrix has a large enough number of rows or columns (the number will depend on screen resolution), only a portion of the matrix will be visible in the matrix form at any one time. When the matrix is larger than the matrix form can contain, the matrix form can be thought of as a viewport onto the matrix. In that case, the vertical and horizontal scroll-bars on the matrix form can be used to show rows and columns that are not being currently displayed.

4. After examining the matrix, close the matrix form to remove it from the desktop. A good practice is to remove from the desktop all matrix forms that are not currently being used. Doing this has two benefits. First, the desktop is kept from becoming too congested, making it easier to navigate among the various objects on the desktop. Second, Windows has the resources to support only a finite, though large, number of objects on the desktop. To check for the percentage of Windows resources that are available at any time, click the resources display in the lower right-hand corner of the desktop. Click this display once more to restore it to showing available memory.

2
Data Analysis

Introduction

This section will introduce a variety of approaches to the analysis of data. The primary focus will be on the application of neural network-based techniques to the tasks of prediction, classification, and function approximation. This section will therefore begin by discussing the following neural network functions that are available in Simulnet:

Network Functions

- *BackProp Network*: Implements a multilayer feedforward network trained using an enhanced form of the error backpropagation learning rule. This network can model the functional relationship distributed among a set of training exemplars. In this way the network can carry out prediction, function approximation, and classification tasks.
- *Genetic Network*: A neural network-genetic algorithm hybrid, in which a genetic algorithm approach is used to evolve a population of feedforward, error backpropagation networks in order to create the optimum network for the particular task at hand. This optimum network can then be applied to prediction, approximation, and classification tasks.
- *Probabilistic Network*: A form of generalized regression neural network using a single-pass algorithm. This function can also abstract a model of the functional relationships embedded within a series of training exemplars. This network can be used for prediction, approximation, and classification.
- *Self-Organizing Map*: An unsupervised vector-quantizer in which an array of network nodes self-organizes to develop decision boundaries between classes of exemplars. The self-organized network can then be applied to the task of classifying novel exemplars.
- *Learning Vector Quantizer*: A supervised vector quantizer network in which an array of network nodes is trained to develop decision boundaries between categories of training exemplars. On the basis of these boundaries the network can then classify novel exemplars.

Other Analytic Functions

In addition to discussing these network-based analytical approaches, this section will also deal with a number of other analysis functions that are available in Simulnet:

- *Fractal Dimension Analysis*: Computes an estimate of the correlation dimension, a measure of the complexity of the dynamical system, a sample of whose behavior is present in the measured data. Correlation dimension can be computed either for a single matrix column, in which case the data is embedded within phase spaces of successively higher dimensionalities, or for a matrix or selected set of matrix columns, in which case phase space embedding is not required. This function features automated computation of both the value of lag and the optimum value of the regression coefficient of correlation integral versus distance.
- *Fourier Analysis*: Carries out an analysis of the frequency components within the data using a Fast-Fourier Transform. This function generates both the complex Fourier components, as well as the frequency spectrum.
- *Eigenvalue Analysis*: Computes the eigenvectors and associated eigenvalues of a matrix. Eigenvectors represent mutually independent linear combinations of the original variables. The amount of the variance in the original matrix that is accounted for by each of these eigenvectors is specified by the size of the corresponding eigenvalues.
- *Analysis of Association*: This category includes three types of analysis; coherence, mutual information, and correlation.
    - *Coherence Analysis*: Provides a measure of linear association between the frequency and phase components in two sets of data. In descriptive terms, coherence is a measure of how closely periodic components in the two sets of data are matched.
    - *Mutual Information Analysis*: Provides a measure of how much information can be gotten about one set of data by making a measurement on a second set of data. In other words, mutual information is an indication of how much the uncertainty about one data set is reduced by a measurement of the second data set. Mutual information can also be considered to be a measure of general, rather than linear, association between two sets of data.
    - *Correlation Analysis*: Along with covariance analysis, provides an index of the extent to which two data sets are linearly related. That is, how much of the behavior of one data set can be accounted for in terms of a linear function of the second data set. While correlation discards information about relative difference in variance, covariance takes differences in variance into consideration.

Each type of analytical approach is intended to examine the data from, to some extent, a unique perspective. Each of these perspectives can in turn be stated in the form of a question that the analyst may be asking of the data. Such a substantive question will be presented for each analytical technique to be discussed.

The Substantive Question

An early decision to be made is whether the data is being analyzed from an exploratory perspective, or from the perspective of an a priori model regarding the process that has been measured or the system whose historical behavior is being examined. If the data analysis is being conducted with an exploratory motive, there may not necessarily be any initial hypothesis or model with respect to the data. If, on the other hand, the analysis is carried out within the context of a model, the model should be capable of providing a question or set of questions for subsequent analysis to address.

More specifically, such a before-hand model should be able to predict that the data will exhibit certain characteristics or constellations of features. Each of these characteristics can in turn be expressed as a corresponding substantive question. The data may then be subjected to one or more analytic probes in order to determine the level of support for each of the questions.

As an example, one technique that will be discussed in this section, Fourier analysis, considers data in terms of periodic features, or frequency components. The substantive question that is addressed by this technique is a straightforward one: What are the relative magnitudes and associated phase angles of the frequency components into which the data can be decomposed? In more substantive terms this question can then be expressed as: What are the relative magnitudes of the periodic features within the data?

The following subsections will each describe an analytic approach, state the objective of carrying out that analysis in the form of a substantive question with respect to the data, and provide examples of how the analysis can be carried out using Simulnet.

Neural Network Analysis

Introduction

By simulating some of the features of biological networks of neurons, artificial neural networks are able to analyze data for patterns, and then make predictions on the basis of those patterns. Significantly, neural networks can do this without the aid of a before-the-fact model of what the data is expected to contain. By not requiring such a model, neural networks can be more flexible, and thus often more powerful, than model-driven analytic techniques.

While neural networks may not require that the data analyst have a specific model or hypothesis about the data, there is always the implied metahypothesis that the data does in fact contain patterns or functional relationships among the sampled variables, which the network is then expected to learn. This point is important for the following reason. Neural networks are voracious learners. Given a finite set of training examples consisting only of random numbers, or noise, a neural network will 'learn' the noise. In other words, the network will learn the pattern represented by the finite set of random numbers. Of course, such learning can not be expected to generalize to any other situation. Nevertheless, the network has learned something, albeit, something that is of limited use.

Artificial neural network models were inspired in part by network theories of how storage of information and learning occur in biological networks of neurons. One such theory is that of Donald Hebb (Hebb, 1949). According to Hebb's model, learning and memory are phenomena which result from the strengthening of the synaptic connections between simultaneously active neurons. In his own words, "When the axon of cell A is near enough to excite a cell B ... A's efficacy, as one of the cells firing B is increased." (Hebb, 1949). Indeed, it has been shown that repeated stimulation of some network of neurons eventually does result in permanent changes in the interconnection strengths between the neurons within this network. The result is that a memory trace has been laid down, or in alternative terms, that learning has occurred. Hebbian learning mechanisms directly or indirectly form the basis of many of the learning mechanisms used to train artificial neural networks.

At a fairly low level, biological neuronal networks are composed of nerve cells, or neurons. Neurons individually perform a relatively straightforward function, as diagrammed in Figure 2.1. Neurons receive signals from other neurons through connections called synapses. A synaptic connection has the important property that it can transmit varying amounts of a signal. In a simple sense, this is not unlike a valve allowing a variable amount of fluid to pass. The strength of a synaptic connection can be modified in response to the activity of the sending and receiving neurons, the principle that is summarized in Hebb's theory.

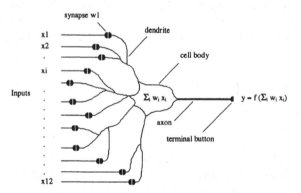

Figure 2.1. The figure shows a highly schematic diagram of a neuron. Input signals x1 through x12 are conducted to the cell body through synapses w1 through w12. The cell body sums the weighted inputs, giving $\Sigma x_i w_i$. When sufficiently large, this weighted sum generates a pulse that travels along the axon away from the cell body. This firing of the neuron can be viewed as the weighted sum passing through a nonlinear transfer function f(.) associated with the axon. Actual neurons can have as many as 10,000 inputs and more than 1 output.

The signal received by a neuron through any one synapse is thus modulated, or weighted, by the strength of that synaptic connection. The neuron then sums all of these weighted input signals. Finally, the sum is passed through what amounts to a nonlinear function: When the value of the sum reaches a threshold, the neuron fires, generating an output pulse. Furthermore, if the sum remains above the threshold the neuron fires repeatedly, generating a train of pulses.

An important property of this pulse train is that the pulse repetition rate, the number of pulses per second, is proportional to the amount by which the sum of weighted inputs exceeds the threshold value. Why then does a neuron exhibit the complex behavior of generating a variable frequency pulse train?

One, and possibly the major, reason is that the sum of weighted inputs represents an analog quantity—a continuous variable. Transmitting such a quantity is prone to the effects of distortion and interference from surrounding activity. One way in which an analog signal can be transmitted with relative immunity to such interference is to convert the analog signal to some type of discrete-values format. One such format is referred to in engineering terms as pulse-code modulation. Pulse-code modulation means simply converting an analog signal, that is to be transmitted over any distance, through a potentially interfering medium into a pulse train whose properties are a function of the value of the analog signal. For example, the pulse-width, or the pulse-repetition rate, can be made proportional to the value of the analog signal. And so, as latterly communications engineers have found that pulse-encoding is an efficient way to transmit an analog signal, it may be speculated that evolutionary pressures have similarly sculpted the transmission mechanism within biological networks to obtain the performance required for the network to function adequately.

A second possible reason for the conversion of the sum of weighted inputs to a pulse train, is to introduce a nonlinear term in the transfer function of the neuron. As will shortly be made clear, the property of nonlinearity is crucial to the ability of networks of artificial neurons to learn arbitrarily complex functions. Without the nonlinearity, artificial neural networks would be restricted to learning tasks that involved only linear function, or classification problems that were linearly separable.

To explain what is meant by the term linearly separable, suppose that we have 20 objects, for example mineral samples. Each sample has a value along each of two dimensions, for example, weight and color. Our task is to learn to separate the 20 objects into 2 groups on the basis of these 2 dimensions. Now the difficulty of this classification task depends on how the 20 samples are arranged along each of the 2 dimensions. We might, for instance, be able to separate the samples by means of a simple linear function—a straight line or decision boundary. In such a case we would say that the samples are linearly separable into the two groups. Figure 2.2a shows an example of a linearly separable classification task. On the other hand, it may be that the samples are so arranged on the two dimensions that they can not be separated into two groups using a linear decision boundary. Rather, we would need a curved—nonlinear—boundary. Figure 2.2b shows a corresponding classification task that, although straightforward, is not linearly separable; there is no way in which the two classes can be separated by a linear decision boundary. In this illustrative example we are dealing with only two dimensions. Correspondingly, we can speak of decision lines. When three dimensions are involved we would be dealing with a decision surface, either flat, corresponding to the straight line in two dimensions, or curved. With more than three dimensions we speak of separating the classes by means of a hyperplane or hypersurface in the space of the dimensions of the objects to be classified.

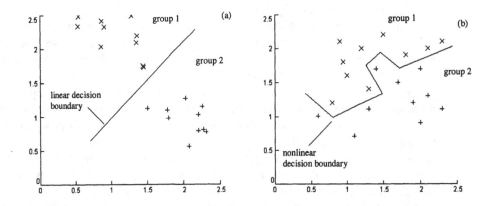

Figure 2.2. (a) Groups 1 and 2 can be classified using a straight line—linear—decision boundary. Such a decision boundary could be generated by a single-layer network. (b) Groups 1 and 2 can no longer be separated by a straight line. A neural network using only a single layer or using only linear nodes would not be able to learn to distinguish these two groups.

An artificial model of neurons, the perceptron, was proposed by Frank Rosenblatt in 1958. The perceptron consisted of modules that bear some degree of correspondence to the functions carried out by neurons. Adjustable weights, analogs of synapses, connected input signals to a summing unit. The output of the summing unit was in turn processed through a nonlinear function—a threshold function that produced an output if its input was above threshold, and no output otherwise. Figure 2.3 shows such a structure, with the nonlinearity generalized to a sigmoidal function. A threshold function can be approximated by choosing an arbitrarily high gain, or steepness, for the sigmoidal function.

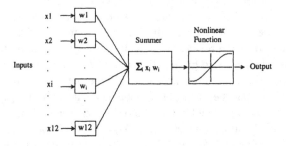

Figure 2.3. The figure shows a block diagram of a perceptron-like artificial neuron. While the perceptron used a binary-valued threshold function, here the nonlinearity is depicted as a general sigmoidal function. Input signals are multiplied by, or weighted by, weights w1 through w12. The weighted inputs are then summed together to produce the sum $\Sigma_i x_i w_i$. This sum is then transformed by the nonlinear function to produce the perceptron output.

Rosenblatt was able to prove an important result about perceptrons; the perceptron convergence theorem. This theorem stated that if a function was computable by a perceptron, the perceptron could be trained to learn that function. This is a very important result in that, even currently, very few neural network models can be proved to be able to learn what they can in principle compute. While this was in itself a seminal step in modeling distributed learning processes, it nevertheless proved to be fundamentally limited. Perceptrons, either singly or connected in single-layer networks, were shown to be limited to computing, and therefore learning, only linear functions. In terms of classification, perceptrons would only be able to learn to distinguish between linearly separable classes. While many real-world problems can be well approximated by such linear function, many others can not. This point was made convincingly by Marvin Minsky and Seymour Papert in their 1969 book *Perceptrons*. In this book the authors pointed out that perceptrons were limited to learning linearly separable classes. In so doing, they succeeded in diverting much research effort away from neural networks. A probable beneficiary of this diversion of research focus was the somewhat complementary approach to machine learning at that time espoused by the artificial intelligence movement, of which Minsky and Papert were exponents. The artificial intelligence approach focused on approaches such as symbol processing and rule structures, rather than the distributed learning perspective of neural networks. Research into neural networks then languished for a number of years. At the same time, it was evident even as Minsky and Papert's book was published that the way out of the impasse for connectionist models of learning using perceptrons was to introduce a level of perceptron nodes not directly connected to either the inputs or outputs of the model—a hidden level. With the addition of such a hidden level, it was known that networks of perceptrons would then be capable of computing any arbitrary function, linear or nonlinear. What was missing at that time, however, was some method, or algorithm, by means of which the weights connecting the input-level and hidden-level perceptrons could be adjusted to allow the network to learn. Intriguingly, Paul Werbos, as part of his 1974 doctoral dissertation, proposed just such an algorithm (Werbos, 1974). His work went largely unnoticed however. It was not until the mid-1980's that the method that Werbos had described was revived, through the efforts of the parallel distributed processing (PDP) research group (McClelland et al., 1986). The work of the PDP group resulted in this algorithm becoming widely known, rekindling interest in research into neural network models. This algorithm, known as error backpropagation, will shortly be discussed in some detail.

Neural networks were inspired by, and developed at least in part, as computer models of the principle of Hebbian learning which underlies learning and memory functions within biological neuronal networks. In broad terms, the brain receives an ongoing stream of sensory information from its environment. On the basis of such sensory information, rules are induced about the behavior and characteristics of that environment. Internal representations, or models, are thus formed of some of the features of the environment. These internal models allow predictions to be made about a future state of affairs. The ability to make such prediction is, in turn, related to an organism's probability of survival. An evolutionary advantage is thus conferred on any such organism that is able in this way to anticipate future conditions based on past experience. Connectionist models such as neural networks have been developed as computational models of such rule-inducing systems. Neural network models are designed to learn the rules or features distributed over a set of examples presented to the network during training.

Thus, in a way that is functionally analogous to the behavior of biological neuronal networks, artificial neural networks learn from examples. These training examples are instances of the functional relationship between a set of independent and dependent variables that the network is expected to learn. A neural network is trained by presenting it sequentially and repeatedly with the set of examples. In the following discussion, network training examples will be referred to as exemplars. In specific terms, the values of the independent and dependent variables of the functional relationship to be learned are represented within an exemplar as a set of numbers. This set of numbers will be referred to as an exemplar vector.

A brief digression will describe what is meant by the term vector. Readers familiar with the concept may choose to go on to the next paragraph. A vector in this context is simply a collection of numbers, each of which is referred to as an element of the vector, that together describe the state of some system. We could, for instance, describe each person within a population in terms of an age-income vector: To each person in the population we would assign a pair of numbers. One of the numbers would represent the person's age in years, the other their annual income in dollars. A particular person might then be associated with the vector {35, 27000}. We could display the age-income information about the members of this population by constructing a graph. To do this we would treat the elements in each vector as the x and y coordinates of a point on an xy graph. We would then have one point on the graph for each person in the population. The resulting plot, referred to as a scatter plot, would be a graphical description of how the members of the population are distributed on the dimensions of age and income. If we wanted a richer description of each person we could extend the vector. We could add elements to the vector to represent height, educational level, blood-pressure, and anything else about a person that could be expressed numerically. Vectors consisting of three elements could be plotted on a three-dimensional graph by treating each vector as the coordinates of a point in three-dimensional space. With an even larger number of dimensions graphing would, of course, become increasingly difficult. The usefulness of vectors extends far beyond the construction of graphical representations of the information they contain. The area of mathematics known as linear algebra involves, in part, the study of analytical techniques for operating on vectors in order to process the information they contain.

To return to the discussion of exemplars, an exemplar vector consists of two sections, the predictor section and the criterion section. The predictor section consists of the values of the set of independent variables of the relationship. The criterion section consists of the values of the set of dependent variables of the relationship. For instance, we may want to train a neural network to predict the behavior of two variables, the value of a particular stock, and whether that stock will likely rise or fall in value in the near future. These two variables would then be the dependent variables of the functional relationship. Furthermore, as the basis on which the network will make its predictions, we may wish to use the values of five different economic indicators. These could include variables such as the value of the domestic currency, unemployment rate, and so on. This group of five variables would then constitute the independent variables of the relationship. The predictor and criterion sections of an exemplar together form a single example of the relationship that the network is asked to learn. Exemplars will be discussed in more detail in a later section.

Network Structure

The following discussion of neural network architecture will use a naming convention that has been proposed in the literature. First of all, the term node will refer to the artificial neurons depicted in Figure 2.3. The proposal has been made that rows of nodes should be referred to as levels, while the blocks of weights interconnecting any two node levels should be referred to as layers. Any glance at the neural network literature will quickly show, however, that many writers still use the term layer to refer to both rows of nodes and to blocks of weights.

The type of neural network implemented in the Simulnet *BackProp Network* function is referred to as a multilayer, feedforward, error backpropagation network. The term multilayer refers to the structure of the network: The network contains at least three levels of nodes, where the term node refers to a simulated neuron. A passive input level simply allows training exemplars to be presented to the network. A hidden level computes nonlinear combinations of the network inputs and acquires, over the course of training, information about some subset of the features distributed within the set of training exemplars. An output level computes nonlinear combinations of the outputs of the hidden nodes. These combinations form the outputs of the network.

The term feedforward refers to the manner in which information is propagated through the network: Information in the form of a training exemplar applied to the network inputs appears, after being processed by the network, at the output nodes of the network. These outputs are not fed back into the inputs, as is the case with some types of networks known as recurrent networks. What is propagated from the outputs back to the inputs is information about the errors that the network has made with regard to the current training exemplar. Error information is propagated back through the network. In an analog of the way in which synaptic strengths are modified in biological neuronal networks, this error is then used to modify the strengths of the interconnections between the network nodes. The training rule that accomplishes this process is correspondingly referred to as error backpropagation. These features will now be discussed in more detail.

Structurally, the network is organized as a number of input nodes, equal to the number of independent variables; and a number of output nodes, equal to the number of dependent variables. Between these input and output node levels is the third set of nodes, the hidden nodes. These nodes are hidden in the sense that they do not directly interact with the outside world. These three groups of nodes will be referred to as the input level, hidden level, and output level of nodes. This structure is diagrammed in Figure 2.4. Linking the input and hidden levels of nodes is a set of connections referred to as the input-to-hidden layer weights. Similarly, connecting the hidden and output levels of nodes are the hidden-to-output layer weights.

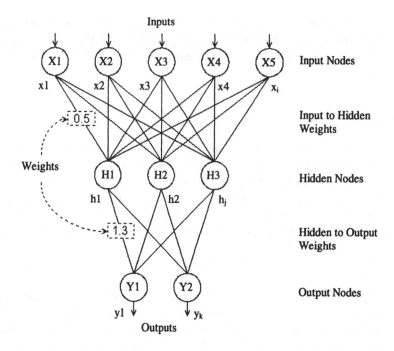

Figure 2.4. This neural network consists of three levels of nodes: An input level of nodes, X_1 through X_5, receives the predictor part of each network training exemplar. The hidden-level nodes, H_1 through H_3, represents, after training, data features that the network has identified from the exemplars. The output-level nodes, Y_1 to Y_2, provide the network's estimate of the outcome associated with each exemplar. Network weights are successively modified as the network learns the relationship distributed within the set of training exemplars. As an example, the weight from node X_1 to node H_1 is shown having a value of 0.5. This means that the output of node X_1 is multiplied by 0.5 before being presented to node H_1. Similarly, the output of node H_1 is multiplied by 1.3, in this example, before being presented to node Y_1.

Each connection, or weight, can vary in strength. Referring to Figure 2.4, a network will consist, in general, of m input nodes, X_1 to X_m; n hidden nodes, H_1 through H_n; and p output nodes, Y_1 through Y_p. The output signal generated by a node is referred to as the activation value of the node. An input node X_i generates an activation value denoted by x_i. A hidden node H_j generates an activation value denoted by h_j. An output node Y_k generates an activation value denoted by y_k. The activation value of input node X_1, for example, is sent to hidden node H_1 after being multiplied by 0.5 (in this example), the value of the weight that connects nodes X_1 and H_1. The activation value of hidden node H_1 is sent to output node Y_1 after being multiplied by 1.3, the value of the weight that connects nodes H_1 and Y_1.

Each hidden node receives inputs from all of the input nodes X_i, with each input being weighted by its own independent weight value w_{ij}. At the hidden node, these weighted inputs are summed. A hidden node H_j thus receives the weighted sum of the values applied to the input nodes $\Sigma_i w_{ij} x_i$ and in turn generates an output value

$$h_j = \phi \left(\Sigma_i w_{ij} x_i \right) \qquad\qquad 1$$

where h_j is the output of hidden node H_j; x_i are the signals applied to, and generated by, the input nodes X_i; w_{ij} are the weights connecting the input nodes to hidden node j; and ϕ is the node activation function that will be discussed below.

It is the presence of the hidden nodes that give neural networks their power to model almost any functional relationship between the independent and dependent variables. In turn, the one property of the hidden nodes that endows them with this ability is their input-output, or activation, function. This activation function is typically a nonlinear relationship, such as the logistic function,

$$\sigma(x) = 2 / (1 + \exp(-gx)) - 1$$

where g is a gain factor that determines how quickly the function reaches its limiting values. This function is graphed in Figure 2.5 (for $g = 1$).

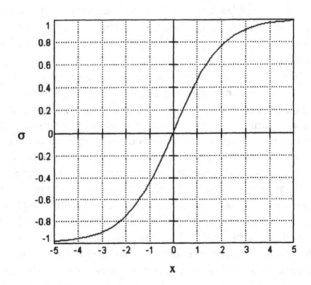

Figure 2.5. The logistic function $\sigma(x)$ is typical of the nonlinear transfer functions used for network hidden-level nodes. The function tends towards -1 for large negative values of x, and towards +1 for large positive values of x.

The x axis values of the graph could represent the combined, or net, inputs to hidden node H_j. This combined input would then be the weighted sum of the signals from the input nodes X_i. The y axis of the graph would then represent the corresponding output from hidden node H_j. In a similar way, each output node receives signals from all of the hidden nodes H_j, with each signal being weighted by its own connection weights w_{jk}. At each output node these weighted signals from the hidden nodes are summed as $\Sigma_j\, w_{jk}\, h_j$. In turn, each output node Y_k generates an output value given by

$$y_k = \phi\,(\,\Sigma_j\, w_{jk}\, h_j\,) \qquad\qquad 2$$

where y_k is the output of hidden node k; h_j are the values generated by the hidden nodes; w_{jk} are the hidden-to-output layer weights connecting the hidden nodes to output node k; and ϕ is an activation function that may be either a nonlinear function, such as the logistic function described above, or the truncated linear function

$$\phi(x) = gx;\ -1/g \le x \le 1/g$$

$$\phi(x) = -1;\ x < 1/g$$

$$\phi(x) = 1;\ x > 1/g$$

where g is the gain factor. Whether an output node has a linear or a nonlinear activation function will depend on the type of learning task required of the network.

When the network learning task involves classification of exemplars into two or more categories, or a pattern classification task, output nodes are usually given a nonlinear activation function. With this type of learning task each output node is generally associated with one of the classes being discriminated, although other configurations are possible. With a sufficient number of hidden nodes and a large enough set of training exemplars, and provided the network does not become trapped in a local minimum (more about these later), the value generated by each output node will be the probability of its respective category. During training, when an exemplar of a particular category is presented, these values will eventually tend towards one for the node representing a particular category, and towards zero for all other nodes. During testing, when a novel exemplar is presented, output nodes will generate values representing the probability that that novel exemplar belongs to each of the categories represented by the output nodes.

When the learning task involves prediction, or function approximation, output nodes are generally given linear activation functions. With this type of task the aim is to represent the sought-after function in terms of a superposition, or addition, of multiple nonlinear combinations of the network inputs. This process is similar to that which is performed by a Fourier analysis: Fourier analysis breaks down a complex function into the superposition of a series of sine and cosine functions. In the case of neural network function approximation, the neural network similarly attempts to represent some complex functional relationship in terms of the nonlinear activation functions of the hidden nodes, typically the logistic function, rather than the sine and cosine functions of Fourier analysis.

For a network with a single hidden level of nodes, the network outputs can be expressed in terms of the network inputs by combining equations 1 and 2:

$$y_k = \phi\,(\,\Sigma_k\, w_{jk}\, \phi\,(\,\Sigma_i\, w_{ij}\, x_i\,)) \qquad\qquad 3$$

The highest level interpretation of equation 3 is that network outputs are nonlinear combinations of network inputs. The job of the network during training is to learn to form these combinations. The network structure, as just described, can be extended to an arbitrary number of hidden layers. It has been shown, however, that a network with even a single hidden level containing a sufficient number of nonlinear hidden nodes, and trained using the error backpropagation learning rule, is capable of acting as a universal function approximator. This result is referred to as the universal approximation theorem (Funahashi, 1989, Hornik, Stinchcombe and White, 1989). That is, a feedforward network with a single hidden level that contains a large enough number of hidden nodes, and trained using the error backpropagation learning rule, can approximate any continuous function to any degree of precision. This property is responsible for the power of multilayer networks, trained using error backpropagation to be able to model complex functions and to perform a wide range of prediction and classification tasks.

Training the Network

A neural network is trained by presenting it with each of the training exemplars, one at a time. An exemplar is presented to the network by feeding each of the input nodes with one of the elements of the predictor section of the exemplar. Thus, each input node receives the value of one of the independent variables that together comprise the predictor section of the exemplar. For each set of predictor values, that is, each exemplar, that is successively presented to the neural network, the network generates a set of output values. These outputs can be considered to be the guesses, or estimates, that the network makes as to what the values of the dependent variables for the particular training exemplar should be. These output values are generated by the network output nodes. Each output node represents one of the dependent variables of the relationship the network is to learn.

At the start of network training the actual outputs of the network will, in general, not be close to the desired outputs for each of the training exemplars. Recall that these desired, or target, outputs are specified in the criterion section of the exemplars. In the process of training the neural network, the network is presented with each training exemplar in turn. The corresponding outputs generated by the network are compared with the target outputs. The difference, the training error for that exemplar, is used as the basis for a scheme that modifies the network weights. The weights are modified according to a rule, typically some form of the error backpropagation rule. This rule is expected, over the course of training, to minimize the difference between the actual outputs of the network and the desired outputs that are coded in the criterion sections of the exemplars.

The Error Backpropagation Algorithm

In order for the neural network to accomplish the goal of learning the functional relationship distributed among the exemplars, the network must develop an internal representation of this relationship. This internal representation is coded in terms of the values of the network weights. These weights are the strengths of the connections among the artificial neurons in the network, the network nodes. In a general way then, these weights are analogs of the strengths of the synaptic connections that exist within biological neural networks, and which are presumed to be the basis for learning in biological networks. Figure 2.4 shows a diagram of a neural network with an input level, a hidden level, and an output level of nodes. The lines shown interconnecting these nodes represent the network weights: Signals from a generating node are multiplied by a factor equal to the value of the weight before being passed to the receiving node.

As described, for each training exemplar that is presented to the network, the network generates a set of output values. These actual outputs are compared with the desired, or target, outputs for that exemplar. By means of an appropriate learning rule, the difference between the actual and target outputs, the training error, is used to modify the weights in the network in a way that is designed to reduce the magnitude of the training error. The particular learning rule used to modify network weight values is typically some variation on the error backpropagation algorithm.

The error backpropagation algorithm first uses the training error to modify the hidden-to-output layer weights. As mentioned, each hidden-to-output layer weight is modified in a direction that is expected to reduce the size of the training error. Furthermore, the size of the weight change is proportional to the contribution that that weight has had in producing the training error. Thus, weights that have had a large effect in determining the current training error are modified by a relatively large amount. Weights that have made a relatively small contribution to the current training error are modified by a correspondingly small amount. After the hidden-to-output layer weights have been updated, the weights in the input-to-hidden layer are modified. These weight changes are made using essentially the same procedure as for the hidden-to-output layer weights. A difference is that while the first set of weight changes used the error levels of the output nodes as the basis of the weight modification, this second set of weight changes uses corresponding error values for the hidden nodes as the basis of the weight change. These error values are again a measure of the extent to which each of the hidden nodes has contributed to the network error.

This sequence of operations thus involves first calculating the training error, the error value of the output nodes; then using this error to successively modify each layer of network weights, starting with the weight layer closest to the network output and ending with the layer closest to the network inputs. The training error thus propagates backwards through the network, from outputs to inputs, in order to update the weights. From this behavior is derived the name of the algorithm, error backpropagation.

The Error Surface

The process of training a neural network can be described using a topographical description of the process that the network goes through in trying to minimize training error. We can conceptualize the information contained within the set of network training exemplars as a landscape in an n-dimensional space. Here, n is equal to the number of variables that are required to fully describe the behavior of the network. This value is approximately equal to the number of weights in our network. This landscape will therefore be referred to as a weight space. With a very simple network of two weights, the landscape may be visualized as a two-dimensional space, a plane. With a larger number of weights, the landscape becomes an unimaginable hyperspace. Nevertheless, the same conditions apply whether there are two or more dimensions. The following description will use an example that can be easily visualized: A simple network with only two weights. This network will have associated with it a two-dimensional weight space, a plane, in which distance along one direction given by the value of one of the weights, and distance along the second direction given by the value of the other weight. In particular, the two weights will be allowed to assume values from zero to one. The weight space thus becomes a plane with length extending from zero to one and width extending from zero to one.

For any combination of values of the two weights, there will be a corresponding value of network training error. In order to represent training error in this description, a third dimension, height, will need to be added to the space. Height, above the two-dimensional plane of possible weight values, will represent the magnitude of the training error. In this example, training error can take on values between zero and one. These values of training error, one value for every combination of values of weights one and two, can be visualized as a surface, and will be referred to as the error surface. The height of the error surface represents network training error. As shown in Figure 2.6, the error surface can have an arbitrarily complex topography. In this representation of the error surface, error varies from zero, or slightly below the lowest point in the surface, to one, or slightly above the highest point in the surface. The topographical features of the error surface are defined by the information distributed among the exemplars that are used to train the network. Typically for real-world network training tasks, the topography of the error surface is unknown. If the shape of the error surface was known it would also be evident what combination of values of the hidden weights would result in the lowest value of network error. Finding this particular combination of weight values is indeed the reason for going through the process of training the neural network.

Now, it is the function of the network to develop values for these weights that will minimize the network training error. In terms of the error landscape, this amounts to saying that the network needs to find the point in the weight space that corresponds to the minimum training error, that is, the lowest point in the error surface. At the start of network training, network weights are generally assigned values drawn at random from some distribution. In certain special cases information may be available about the problem to be solved, in which case it may be possible to assign weight values more knowledgeably. In effect, the network can, in such cases, be given a head start in its training. In any event, this set of initial random weight assignments defines a point on the error surface, the starting point for the network training. Training the network can then be pictured as proceeding along some complex path along the error surface, from the starting point to—it is hoped—the lowest point on the surface. This lowest point corresponds to the point of minimum training error.

This training process would be straightforward were it not for the fact that with virtually all network training data the error surface contains obstructions and other terrain features that make the path from start point to end point a highly convoluted one. The landscape contains the equivalent of features that in two-dimensional space might be described as valleys, hills, ravines, and flat plains.

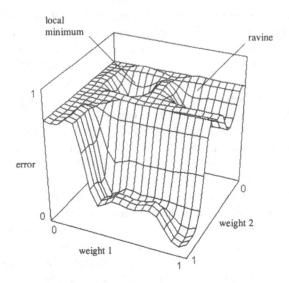

Figure 2.6. The weight space of the range of possible values of weights 1 and 2, and the corresponding value of network training error, defines an error surface. This error surface gives the value of training error for every combination of values of weight 1 and weight 2. A global minimum of the error surface can be seen in the neighborhood of weight 1 = 1 and weight 2 = 1. Two local minima can be seen. One occurs near weight 1 = 0.3 and weight 2 = 0.3. The second local minimum is a shallow ravine in the area of weight 1= 0.3 and weight 2 = 0.6 to 1.

With the image of the error surface in mind, the process of training the network can be described as starting from a point on the error surface defined by the initial values of the network weights, and trying to find the point of minimum training error. In order to do this the network must successfully navigate the various obstructing features within the landscape. The strategy that the network uses, given by the error backpropagation algorithm, is as follows: At whatever point the network finds itself on the error surface, it examines the slope of the surface along the various dimensions of the weight space (recall that there are two dimensions in our example of a two-weight network). The network then moves from where it is to a new point by moving along each of these dimensions an amount that is proportional to the slope of the surface. Thus if, for example, the slope of the error surface along the weight 1 dimension were large and the slope along the weight 2 dimension were small, the network would move from where it is to a new point by taking a correspondingly large step along the weight 1 dimension of the error surface and a correspondingly small step along the weight 2 dimension. The direction of the steps is always in the downward direction. Phrased in this way, the behavior of the error backpropagation algorithm should be seen to be eminently reasonable. A strategy of 'the greater the slope, the larger the step' should eventually succeed in getting to the lowest point in the error surface.

Complications arise, as hinted earlier, because of features in the error surface such as flat planes and low points that are, however, not the lowest points on the surface. The problem with flat regions is that should the network find itself in such an area, it would not be able to compute a slope in any direction: A flat region has a local value of slope of zero. Experience shows that regions that are nearly flat, with correspondingly very small values of slope, do occur. Such regions can pose a real problem for the network. With real data, the chance of a truly flat region in weight space is vanishingly small. This is related to the fact that, as there is an infinity of real numbers in any given range, the probability of obtaining any one real number is correspondingly zero. Of concern in practice are regions that are nearly flat. Such regions do not stop the network from proceeding, but they can slow it down, sometimes to the point that the network becomes effectively untrainable with the given training data.

The second type of topographical feature that can be problematic for a network is a low point that is not the lowest point in the error surface. Such points are termed local minima to distinguish them from the lowest point in the error surface, which is referred to as the global minimum. The potential problem of local minima is, however, also quite rare with real data. In order for a weight space to contain a true local minimum, the value of error must rise along all dimensions of the weight space. With a two-dimensional weight space as depicted in Figure 2.6, local minima are easy to construct. The more dimensions there are in the weight space, however, the more opportunities there are for the network to escape from the minimum. While this potential problem of local minima has been used to support the argument that neural networks using back propagation are limited in their capabilities, such local minima appear to be rare in practice. Additionally, if a set of training exemplars did appear to generate a local minimum in the error surface, this feature could probably be eliminated by adding one more dimension to the weight space, by adding even one more weight.

Computing the Weight Changes

Error Derivatives

As described above, the error backpropagation algorithm computes the size of the weight changes by finding the slope of the error surface with respect to each weight dimension of the weight space. To put it another way, the size of the change in a weight w_{ij}, Δw_{ij}, is determined by finding out how sensitive the current training error is to changes in the sizes of the network weight. The magnitude of this sensitivity is given by the slope of the error surface with respect to each of the weight dimensions. This quantity is referred to as the error derivative, $\partial E/\partial w_{ij}$ for the weight. The size of the weight change is then

$$\Delta w_{ij} \propto -\partial E/\partial w_{ij}$$

Learning Rate

The error derivative $\partial E/\partial w_{ij}$ is only one of the factors that determines the actual amount by which the value of a weight is modified. A second factor is learning rate, η. The magnitude of the weight change is equal to the product of these two factors, error derivative and learning rate:

$$\Delta w_{ij} = \eta \cdot (-\partial E / \partial w_{ij})$$

Learning rate is the factor that determines the size of the steps that the network takes in navigating through the weight space in order to minimize the magnitude of the training error. The size of the learning rate parameter can be set by the user. As a rule, the higher the value of learning rate, the faster the network will locate the point in the weight space that corresponds to an, at least locally, low training error.

There is, however, a penalty for using a value of learning rate that is too large: Network training may proceed chaotically and may not even converge to a low value. In terms of navigating through the weight space, if the steps that the network takes are too large, any one of a number of undesirable effects may occur. The network may continually overshoot the point of minimum error, never reaching it, but bouncing back and forth around it. Still another possibility is that the network may begin to interact unpredictably with the terrain features of the weight space: In a manner of speaking, the network may start bouncing off the wall of landscape features such as ravines and hills. The result will be classically chaotic behavior. The network is a nonlinear system, and over different ranges of its parameters, it may exhibit the different types of behavior that have been associated with nonlinear systems: point attractors, limit cycles of various periodicities, chaotic behavior, and intermittency.

In contrast, too low a value of learning rate will result in a very slow rate of convergence to a low training error. The optimum value of learning rate is very much dependent on the nature of the network training data since this data determines the topography of the weight space. As a matter of principle, the error backpropagation algorithm only guarantees convergence to a training error minimum in the limit of an arbitrarily small learning rate. Generally, it is only through experimentation and experience that the best learning rate for a given set of training exemplars can be found. Since the error backpropagation algorithm was first proposed, many modifications and enhancements have been suggested to improve the algorithm's convergence characteristics.

Momentum

One of the most straightforward enhancements is the addition of another user-specifiable parameter referred to as momentum, symbolized by α. The function of momentum is to increase the size of a step when the direction, in the weight space, is the same as the direction of the previous step. Conversely, momentum decreases the size of a step when the directions of the current and previous steps are not the same. The idea is that if the network is making good progress in the current direction then it should keep going forward with an even larger step. If, however, conditions have changed between the previous and current steps, then take a correspondingly small step. As with learning rate however, there are practical limits to the size of momentum. If momentum is too large then the network may continually overshoot the point of minimum error. Even with more moderate values of momentum, the network may zigzag across the landscape, thus extending training time. The computation of the weight changes considering both learning rate and momentum is given by the relationship

> weight change = (learning rate · error derivative) + (momentum · previous weight change)

$$\Delta w_{ij} = \eta \cdot (-\partial E/\partial w_{ij}) + \alpha \cdot \Delta w_{ij} \text{ (previous)}$$

Local Error

For any weight w_{ij} connecting the output of node X_i with an input to node H_j its error derivative $\partial E/\partial w_{ij}$ is computed as follows: The error derivative is the product of the local error, also referred to as local gradient δ_j, for a node H_j, and the output of node X_i, x_i, and is computed as

$$\partial E/\partial w_{ij} = -\delta_j \cdot x_i$$

The local error δ_k for an output node is easily computed, since the target value for output nodes are known:

$$\delta_k = E_k, \text{ for linear output nodes}$$

$$\delta_k = E_k \cdot y_k \cdot (1 - y_k), \text{ for logistic output nodes}$$

The local error for a hidden node is more difficult to compute since target values for hidden nodes are not known. Instead, a result from calculus allows the local error for a hidden node to be computed using the local errors for the output nodes together with the hidden-to-output weights:

$$\delta_j = h_j \cdot (1 - h_j) \cdot \Sigma_k (\delta_k \cdot w_{jk})$$

Putting these ideas together, the size of the weight change is computed using the relation

$$\Delta w_{ij} = \eta \cdot \delta_j \cdot x_i + \alpha \cdot \Delta w_{ij} \text{ (previous)}$$

Variable Learning Rate

Among the many enhancements to the error backpropagation algorithm that have been proposed, one involves the use of a learning rate that can be different for each network weight. Essentially, each weight has its own value of learning rate. This learning rate is modified during training using various types of heuristics aimed at improving the convergence behavior of the algorithm (for example, Silva and Almeida, 1990, Vogl et al., 1988). All of these heuristics aim to tailor the size of the step that the network is to take to the features of the error landscape in the vicinity of the current position.

The Error Backpropagation Algorithm

The error backpropagation algorithm can now be described in somewhat more detail. For this description it will be assumed that the network is structured as in Figure 2.4, with an input level of nodes, a single hidden level, and an output level. The algorithm can be extended to more complex network configurations, for example networks with more than a single hidden node level.

The Algorithm

1. Initialize network weight values
2. Repeat the following steps until some criterion is reached:
   2.1 For each training exemplar:
      2.1.1 Do a Forward Pass
      2.1.2 Do a Back Pass
   2.2 Update Weights
   2.3 Test network generalization
3.   Run the trained network

The three component functions, *Forward Pass*, *Back Pass*, and *Update Weights*, will now be described in more detail.

Forward Pass

This function uses the values of the network inputs to compute the values of the hidden nodes, and then uses the values of the hidden nodes to compute the values of the output nodes.

For each hidden node j:

Compute the values of the hidden nodes:

$$h_j = 2 / (1 + \exp(-\Sigma_i (x_i \cdot w_{ij}))) - 1$$

where h_j is the j-th hidden node, x_i is the i-th input node, and w_{ij} is the weight connecting input node i and hidden node j.

In words: to find the value of a hidden node, compute the sum of the products of the values of all input nodes, and the values of the input-to-hidden layer weights. Then, transform this sum using the logistic function $f(x) = 2 / (1 + \exp(-x)) - 1$

For each output node k:

Compute the values of the output nodes

$$y_k = 2 / (1 + \exp(\Sigma_j (h_j \cdot w_{jk}))) - 1$$

where y_k is the k-th output node, and w_{jk} is the weight connecting hidden node j and output node k.

In words: to find the value of an output node, compute the sum of the products of the values of all hidden nodes and the values of the hidden-to-output layer weights. Then, transform this sum using the logistic function defined as above.

Back Pass

Compute error derivatives for each network hidden and output node. These error derivatives can be thought of as a measure of the extent to which a node is implicated in generating the current network error. The larger the error derivative for a node, the greater the extent to which the node is responsible for the error. The error derivative for an output node Y_k, denoted by δ_k, is computed as the partial derivative of the error E_k associated with the node, with respect to the net input to the node, net_k. This partial derivative is written as $\delta_k = \partial E_k / \partial net_k$. The error derivative for a hidden node H_j, denoted by δ_j, is computed as the partial derivative of the error E_j associated with the node, with respect to the net input to the node, net_j. This partial derivative is written as $\delta_j = \partial E_j / \partial net_j$. This computation is carried out starting with the output nodes, and then proceeding to the hidden nodes.

For each output node Y_k compute its error derivative δ_k as follows:

First, compute the raw error for output node Y_k, as the difference between the activation value or output of the node, y_k, and the target value for the node, t_k: $E_k = t_k - y_k$. The target value is the value that node Y_k is being trained to achieve, and is the value of the k-th dependent variables in a network training exemplar.

Then, if output nodes are linear, compute $\delta_k = E_k$

If output nodes are nonlinear (logistic), compute $\delta_k = E_k \cdot y_k \cdot (1 - y_k)$.

For each hidden node H_j compute its error derivative δ_j as follows:

Since target value for hidden nodes are not available, the error derivative for a hidden node can be computed using the error derivatives for the output nodes together with the hidden-to-output layer weights:

$$\delta_j = h_j \cdot (1 - h_j) \cdot \Sigma_k (\delta_k \cdot w_{jk})$$

Again, the significance of these error derivatives is that they indicate the extent to which network error is sensitive to changes in the weights associated with each node. In alternative terms, the error derivative for a node indicates the slope of the error surface in the directions corresponding to those weights. These error derivatives are stored, and used later by the *Update Weights* function to actually compute the magnitude of the weight change.

Update Weights

Compute and record the size of the weight change for each network weight; change the value of each weight by the amount of the weight change. This operation is carried out first on the input-to-hidden layer weights, and then on the hidden-to-output layer weights.

For each input-to-hidden layer weight w_{ij} :

First, compute the size of the weight change

$$\Delta w_{ij} = \eta \, \delta_j \, x_i + \alpha \, \Delta w_{ij}'$$

where Δw_{ij} is the size of the weight change, δ_j is the error derivative of the j-th hidden node, x_i is the activation value of the i-th input node, η is a parameter that determines the size of the step taken by the algorithm in attempting to reduce network error, α is a parameter referred to as momentum that determines the effect that the previous weight change will have on the current weight change, and $\Delta w_{ij}'$ is the size of the previous change in value for the weight. In this expression, the term $\delta_j x_i$ is the error derivative described earlier, the slope of the error surface with respect to weight w_{ij}.

Then, compute the new weight value as $w_{ij} = w_{ij} + \Delta w_{ij}$

For each hidden-to-output layer weight w_{jk} :

First, compute the size of the weight change

$$\Delta w_{jk} = \eta \, \delta_k \, h_j + \alpha \, \Delta w_{jk}'$$

where Δw_{jk} is the size of the weight change, δ_k is the error derivative of the k-th output node, h_j is the activation value of the j-th hidden node, and $\Delta w_{jk}'$ is the size of the previous change in value for the weight. The term $\delta_k h_j$ is the error derivative for weight w_{jk}.

Then, compute new weight value as $w_{jk} = w_{jk} + \Delta w_{jk}$.

These steps are illustrated in Figure 2.7. This figure shows the forward propagation of node activations through the network, from input nodes to output nodes, and a concurrent back propagation of the influence of network errors through the network, from output nodes to input nodes.

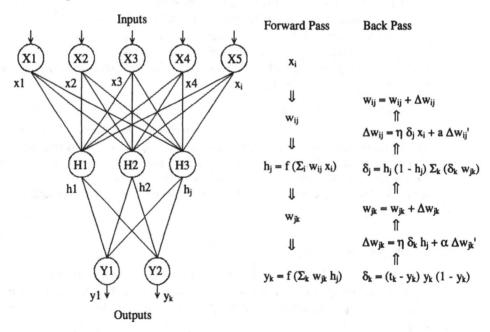

Figure 2.7. In this multilayer, feedforward, error backpropagation network, activations propagate in a forward direction through the network, from inputs to outputs. Concurrently, errors propagate through the network in a backward direction, from outputs back to inputs.

Batch-Mode versus Exemplar-Mode Training

Updating of network weights can be performed after each training exemplar is presented to the network, as described earlier. This schedule of weight updating will be referred to as exemplar-mode training. An advantage of exemplar-mode training is that training error will generally begin to converge to a low value relatively quickly. A disadvantage of this training schedule is that the order in which exemplars are presented may have an effect on the training process. With exemplar-mode training, training exemplars must be presented to the network in a randomized way in order for the network to learn effectively. If this is not done, then it may happen that exemplars are grouped according to features that are present in some exemplars but not in others. If this should happen, the network may not learn these features effectively. Essentially, the network may learn these features as the subgroup of training exemplars containing those features is presented. The network may then, to some extent, unlearn those features as other subgroups of exemplars, not containing those features, are presented. Complete randomization of exemplars may not even be possible: Typically, it will be difficult for the network user to identify the features in the exemplars that are relevant to the functional relationship that the network is expected to learn.

An alternative to this schedule of training is referred to as batch-mode training. With batch-mode training, updating of network weights is carried out only after the entire set of training exemplars has been presented to the network. As each exemplar is presented, partial results are generated and stored. These partial results are combined after all training exemplars have been presented, and are then used to update the network weights. A significant advantage of batch-mode training is that the order in which individual training exemplars are presented is irrelevant. The order of presentation does not need to be randomized as it does with exemplar-mode training. The *BackProp Network* function in Simulnet uses a batch-mode training schedule.

Cross-Validation

Over the course of the training, the network can be periodically tested by presenting it with one or more test exemplars. This procedure, referred to as cross-validation, is essentially one of validating the information that the network has abstracted to see if the network can indeed generalize what it has learned to novel exemplars. The terms cross-validation and testing will be used interchangeably. Test exemplars are constructed in exactly the same way as training exemplars. Typically, a random subset of the entire set of training exemplars is withheld from the training set and then used for testing, or more precisely cross-validating, the network. During testing the network does not use the criterion portion of each testing exemplar to modify network weights. In other words, when a testing exemplar is presented to the network, the network does not train. Instead, during testing the network only generates a set of one or more outputs that represents its guess as to what outcomes are associated with each of the testing exemplars. By examining these outcomes, and seeing how close they are to the target outcomes as coded in the criterion portion of the testing exemplars, we can evaluate the ability of the network to generalize what it has learned to novel exemplars. In other words, we can determine the extent to which the model that the neural network has abstracted from the information distributed among the training exemplars, fits the novel testing exemplars. The question to be answered is the following: Is the model that the network has abstracted one that is general enough to account for the testing exemplars on which the network has not been trained? Or is it a model that, while accounting for the training exemplars quite accurately, does not generalize well to the testing exemplars?

It is the potential ability of the network to generalize from the information in the set of training exemplars, to novel information on which the network has not been trained, that is one of the most useful features of neural networks. The extent to which the network is able to generalize can be quantified by calculating, for each exemplar, a quantity referred to as the testing error. Testing error is the difference between the actual network outcomes and the target outcomes in the testing set. The size of this testing error is an indication of how well the network is able to generalize what it has learned in the training phase to novel exemplars. When testing error is seen to have reached a minimum, the network is considered to have been optimally trained. At this point, the network can be put to work by presenting it with a set of exemplars for which there are no known outcomes. The network will now produce an output for each of these unknown exemplars. This output is the network's predicted outcome for these exemplars. Figure 2.8(a) shows training and testing error decreasing over a number of passes through the network training exemplars.

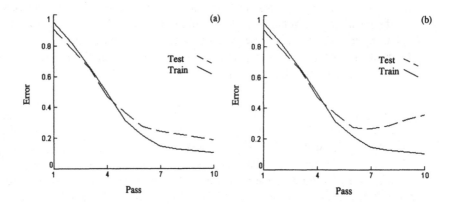

Figure 2.8. (a) The behavior of training and testing error is shown for ten passes through the training exemplars. Training error and testing (validation) error both decrease monotonically to a minimum value. (b) Training error decreases monotonically over the ten passes through the training exemplars. Testing (validation) error reaches a minimum at pass 7, and then starts to rise again. Continuing to train the network beyond pass 7 results in overtraining of the network. In this condition the network loses some of its ability to generalize what it has learned to novel exemplars, such as the exemplars presented to it during testing.

As a practical matter, testing is generally performed at some regular interval such as, for example, once every five training iterations. In other words, for every five sets of training exemplars, one set of test exemplar is presented to the network. As training and testing proceed along such a schedule, testing error should decrease from some starting value, demonstrating that the network has been able to capture, in terms of the values of the network weights, significant features of the relationship between the predictor and criterion variables represented in the training exemplars.

Sometimes however, after some amount of training, testing error will be seen to reverse its falling trend, and to begin to increase. This situation is illustrated in Figure 2.8(b). Testing error thus typically reaches a minimum value at some point during network training. The number of iterations corresponding to the minimum testing error represents the point at which the network has been optimally trained. That is, when testing error has reached a minimum the network is best able to generalize what it has learned from the training exemplars to novel exemplars.

Continuing to train the network beyond this point has the effect that the network attempts to 'overfit' the training data: The network attempts to account for features in the predictor variables that are of progressively lesser significance, and which indeed may represent information unrelated to the sought-after relationship. Such features can be considered to be noise in the context of the pattern being sought.

To restate, overtraining the network, training it beyond the point of minimum testing error, results in the network attempting to account for what can be considered to be noise in the data. Significantly, in doing so the network loses some if its ability to correctly generalize its learning from the training exemplars to the testing exemplars. In this situation, the network has developed an internal representation of the *particular* exemplars that are being used to train the network. This is not what we generally want when training a neural network. What we want is for the network to develop an internal representation, or model, which captures information that is distributed across all of, or at least a significant portion of, the training set, rather than information about the particular exemplars that are being presented to the network. Such an internal model will then represent the features of the training set that are distributed across the exemplars, rather than representing those features that are characteristic of the particular training exemplars. When the network has been trained in this way, the network will demonstrate the lowest possible testing error. The upward turn in the testing error curve when the network is trained past this optimal point (Figure 2.8b) is a signal to the network user that the network is beginning to look at information in the training set that does not generalize to the testing exemplars.

There is still more information that can be gotten from the shape of the testing-error curve. So far, we have described how the point of optimal training is signaled by the minimum of the testing-error curve. There is a second piece of information to be gotten from the testing-error curve. It may turn out that testing error falls, but never again rises as training proceeds. This behavior may indicate that the network contains too few hidden nodes. On the other hand, it may indicate that the number of hidden nodes in the network is actually the optimal number. We cannot know which of these situations is the case, however, until we try increasing the number of hidden nodes, and then retraining the network. Suppose that we increase the number of hidden units by one unit, restart training, and find that the testing-error curve now does show a clear minimum, by falling initially and then rising again as training proceeds. We would then conclude that the network probably had the correct number of hidden nodes before the extra node was added.

At the same time, we generally do not want to increase the number of hidden nodes above the optimal number. As mentioned earlier, with more than the optimal number of hidden nodes the network becomes prone to overtraining. Now, we do have the option of stopping training just as the testing error curve starts to turns up. The time required to arrive at this point however increases with more hidden units. With more hidden units there are more network nodes and weights that need to be trained, and thus a longer training time. Angus (1991) suggests that an effective procedure is to begin with, and 'grow' a network with a small number of hidden-level nodes, rather than to prune a large network. Since the network is thus initially small, such a procedure allows network performance to be assessed in a smaller amount of training time.

Many variations on the basic error backpropagation training procedure have been suggested. Sietsma and Dow (1991), for example, have pointed out that adding noise to training data may sometimes improve the subsequent ability of the network to generalize. In particular, they found that training with noisy exemplars improved a network's ability to classify both noisy exemplars of the trained-on classes, as well as exemplars of classes not present in the training data set.

Network Performance

When we observe, by monitoring the performance of the network during validation, that the network is able to generalize its learning to novel exemplars, we can use the network to generate outcomes for completely novel exemplars for which there is no outcome or criterion information. This last phase of the network operation is sometimes referred to as running the network. To recap then, network operation involves three stages or phases. The first stage is the training phase during which the network acquires the functional relationship distributed among the network training exemplars. In this stage the network is presented with a sequence of training exemplars, each consisting of the values of a set of independent variables, and the values of the corresponding set of dependent variables— the variables whose values the network is to learn to predict. The second stage, which in practice is interleaved with the training phase, is the testing, or validation, phase in which the network demonstrates its ability to generalize its learning to new exemplars. Finally in the third stage, the running phase, the network is put to work to generate estimates of the values of dependent variables given exemplars containing only the values of the independent variables.

Thus, after the neural network has been trained, the network can be used to generate predicted outcomes for novel exemplars for which no dependent variable information is available. As a practical matter, such exemplars consist of the values of predictor (independent) variables only, with the values of the criterion (dependent) variables set to zero (to act as a place-holder in the network data matrix). In this situation, the neural network is asked to apply the model that it has abstracted from the training exemplars to these novel exemplars.

The quality of the predictions made by the neural network is determined by many factors. These include the number of training exemplars relative to the number of predictor variables, the number of training exemplars relative to the number of network weights, the number of hidden nodes, and the number of hidden layers.

Number of Training Exemplars

One consideration regarding the size of the training set in determining the performance of a neural network is the number of training exemplars relative to the number of predictor variables. When the number of predictor variables, and thus the number of network inputs, is greater than the number of training exemplars, the relationships within the training data can be adequately modeled by a linear function. In particular, when a network is being used for pattern classification, having more inputs than training exemplars means that the classes within the training data can be separated by linear, straight-line, decision boundaries (Lisboa, Mehridehnavi and Martin, 1994). Essentially, in this situation the power of the network to develop nonlinear boundaries, or in alternative terms, to model the data in terms of nonlinear functions, is not being used. More traditional and inherently less-powerful statistical methods may be used instead.

As an analogy, suppose that we were being asked to decide how to distinguish between two sets of pictures. Suppose also that we only had a single pair of nonidentical pictures upon which to base this decision. If the pictures were detailed enough, we could certainly develop a simple and straightforward decision on the basis of whatever particular differences there would be between the two pictures. This situation would be analogous to needing only a simple linear rule or decision boundary to separate the two classes. Suppose on the other hand that we had a large number of such pairs of pictures. Our task would now be much more difficult: We would not be able to develop a decision on the basis of particular differences between the pictures in each pair. Rather, we would have to develop a decision on the basis of features that were characteristic of groups of pictures, rather than features that were characteristic of individual pictures. Importantly, we could find that these discriminating features interacted. The effectiveness of particular features as discriminators might depend on the degree to which other features were present. This more complex discrimination task would amount to developing a nonlinear—because of the interactions among the features—decision function or rule.

The heuristic that can be abstracted from this phenomenon is that the number of training exemplars must be greater than the number of network inputs if the network is to develop arbitrarily complex models of the data. In other words, if we want the network to learn complex relationships within the training data, we need more training exemplars than the number of predictor, or independent, variables in each example. If the number of training exemplars is less than the number of independent variables in an exemplar then the network does not need to, and therefore generally will not, develop a nonlinear model to account for the data.

A second consideration, with respect to the size of the training set affecting network testing performance, is the available number of potentially noisy training exemplars for a given size of network. It has been suggested (for example, Baum and Haussler, 1989) that for a single hidden-level network the worst-case testing error, and therefore prediction performance, can be approximately estimated by the relation:

testing error ≈ number of network weights / number of training exemplars

Thus, for a set of 1,000 training exemplars, and a network with 100 weights, the error during testing can be as high as 100/1,000, or 10%. Note that this is a worst-case figure; experience suggests that on average testing error will be smaller than this value. According to this relationship, for a given network size, the larger the number of training exemplars the smaller we can expect the testing error to be. The number of network weights is a simple function of the number of nodes and layers comprising the network:

$$\text{Number of network weights} = (n_i + n_o)\, n_h + (n_l - 1)\, n_h^2$$

where n_i = number of input nodes , n_o = number of output nodes , n_h = number of hidden nodes and n_l = number of hidden layers. The number of input and output nodes is generally fixed by the definition of the problem. The number of hidden nodes and hidden layers however can be varied, and we see that the number of network weights increases linearly as the number of hidden nodes, and as the square of the number of hidden layers.

This point is another argument in support of using the smallest possible number of hidden nodes. Using the minimal number of hidden nodes and hidden layers results in not only the fastest training time to the point of optimal training, but also the reasonable expectation that this point will correspond to the lowest possible testing error.

Number of Hidden Nodes

The estimation of the number of hidden nodes is a key task that must be faced by a user applying a neural network to a new problem area. One approach, discussed above, is to closely monitor testing performance as the network is being trained. To begin, some initial estimate of the appropriate number of hidden nodes is made. A possible strategy is to start with a small number of hidden nodes, for example, two. Network training is then started. When validation error is seen to have stopped decreasing, training is halted. Next, the number of hidden nodes is increased by one, network weights are reinitialized, and training is restarted. Validation error is again monitored with training halted when validation error is seen to have reached a minimum. This process of adding a hidden neuron and restarting training is repeated until validation error is observed to bottom out and to start to increase. The network will now have the optimum, or just over the optimum, number of hidden nodes. In sum, if the number of hidden nodes is too small, validation error will decrease to some value but will not show an upturn as training is continued. For some sufficiently large number of hidden nodes, validation error will be seen to reach a minimum and then show an upturn. At this point the training process can be halted and the network can be considered to be optimally trained.

In order to simplify the task of choosing the number of hidden nodes, Simulnet provides a function that will compute an estimate of the optimum number of hidden nodes for a network configuration consisting of a single hidden level. This estimate involves a heuristic based on the number of significant eigenvalues in the training matrix. As well,

Simulnet provides a function that will estimate the optimum values of the input-to-hidden layer weight values, again, for a network with a single hidden level. These functions are available on both the Simulnet *BackProp Network* and *Genetic Network* functions.

Number of Hidden Levels

As mentioned earlier, the universal approximation theorem states that a network with a single hidden level containing a sufficient number of nodes can approximate any continuous function to any degree of accuracy (Funahashi, 1989, Hornik et al., 1989). Thornton (1992) points out however, that this does not imply that a single hidden level is necessarily the most efficient configuration for all problems. Experience gained through applying neural networks to a variety of problem types suggests that it is generally useful to start with a small number of hidden nodes and hidden layers. The number of hidden nodes and the number of hidden layers can then be systematically varied when first exploring a particular problem. It is in this way that the network user may develop some intuition about the effect of these variables on network performance.

While a network with a single hidden level may in principle be sufficient to model any functional relationship, there can in practice be drawbacks to such an architecture. The first is that the single hidden level may need to contain a large number of hidden nodes. This in turn would make network training computationally intensive, and would then slow the rate at which the network learns. A second potential drawback is that hidden nodes in the single hidden level necessarily interact: As some nodes come to respond more closely to a particular set of local data features, these nodes may affect the response characteristics of other hidden nodes. The result could be that the response of these other hidden nodes may become less accurate with respect to their features. The overall result is that it may be difficult to have each of the hidden nodes represent their corresponding subgroups of data features accurately, at least in a reasonable amount of training time.

How then can the hidden nodes be to some extent uncoupled from each other? In other words, how can the network be designed so that if some nodes learn some of the data features well, the learning ability of other nodes will not be disrupted? One way of accomplishing this is to add a second hidden level (Chester, 1990; Funahashi, 1989). The nodes in the first hidden level will then be relatively freer to extract local features in the data. Nodes in the second hidden level will then be able to complement this process by extracting data features of a more global nature. The result is that the network is allowed to extract data features in a hierarchical fashion: Lower level features in the first hidden node level, higher level features in the second hidden level. This behavior of the network arises from the fact that nodes in the second hidden level receive inputs from nodes in the first hidden level. Nodes in the second hidden level are therefore able to combine information about sets of local features that have been abstracted by the first hidden level. In this way, the second hidden-level nodes are able to construct higher-level concepts about the features in the training data. It may even be the case that, for some types of problems characterized by hierarchically highly structured data features, a network architecture containing a still larger number of hidden layers, each with a small number of hidden nodes, might be appropriate.

In any event, a network architecture with two or more hidden layers would be an appropriate choice when the possibility exists that the function to be approximated by the network is discontinuous. A discontinuous function might be informally described as one that is characterized by at least some features whose values are more or less independent across different regions of the function; the function might, for example, have values of slope that are independent of each other in two different parts of the function. According to the universal approximation theorem, such a discontinuous function can in principle be approximated by a continuous function of sufficiently high order. Such a function could be developed by a network with a sufficient number of hidden nodes in a single level. Again, this approach might require so many hidden nodes that training time could grow to become impracticably long. A more efficient alternative with such a data set would be to use two hidden layers. With two hidden layers, nodes in the first hidden level would act to some extent independently in abstracting the independent features of the discontinuous function. These features could then be combined by the nodes in the second hidden level.

Training and Testing Matrices

Network training and testing are carried out by presenting the network with examples of the relationship that the network is expected to acquire, the training exemplars. The set of available exemplars is in practice divided into two groups. One group is combined to form the training matrix. The second group is combined to form the testing matrix. This section will discuss practical aspects of the creation of training and testing matrices.

Training and testing matrices are both organized in the same way; each matrix row contains one exemplar, one example of the relationship that the network will be asked to learn, or on which the network will be tested. As explained earlier, each of these exemplars is composed of two sections. The first, predictor, section contains the values of the independent variables of the functional relationship the network is expected to learn. If there are n predictor variables, the first n elements of each exemplar, and thus the first n columns of the file, contain the predictor values. The second part of each exemplar is the criterion, or outcome, section containing the values of the dependent variables of the relationship. If there are m criterion variables, the next m elements of each exemplar, and thus the next m columns of the file, contain the criterion values. The total number of columns in a network training and testing matrix is therefore n + m. As an example, suppose the network training task involves five independent (predictor) variables, and two dependent (criterion) variables in each exemplar. Suppose further that 50 such exemplars are available for training the network, and 25 exemplars for testing or validating the network. The network training matrix will then be organized as 50 rows of 5 + 2 = 7 columns. The testing matrix will be organized as 25 rows of 7 columns.

Typically, training and testing sets are created by randomly segregating the available exemplars into two groups, with a subset of the exemplars used for training and the remaining exemplars reserved for testing. The set of testing exemplars is formed by repeatedly drawing exemplars at random and without replacement from the set of training exemplars until the required number of testing exemplars has been drawn. These exemplars then form the testing set, while the remaining exemplars form the training set.

Jackknifed Classification

With a limited number of available exemplars an alternative training and testing procedure is sometimes adopted. The set of training and test exemplars is created as follows:

- A single exemplar is removed from the entire data set.
- The remaining exemplars become the training set, while the single removed exemplar becomes the testing set.
- The network is trained on all exemplars in the training set, and then tested on the single test exemplar.
- Testing error is recorded.

This process is repeated for each exemplar in the training set in turn, removing one exemplar at a time to be the testing exemplar. The result is a set of testing errors, one for each of the exemplars in the training set, as that exemplar was in turn used for testing. An overall testing error can then be calculated by combining the individual testing errors, for example by finding the average of the testing errors. This procedure is referred to as the *leave-one-out* method. This method is sometimes loosely referred to as *jackknifing*, or jackknifed classification. Strictly speaking, these two labels refer to identical procedures used for different purposes: The leave-one-out method is used to generate an estimate of the predictive ability of a model. This is the goal in the present case. Jackknifing on the other hand is a technique that is used to estimate performance characteristics, such as bias and variance, of a statistical estimator. Having made the distinction between the jackknifing and the leave-one-out method, we will nevertheless use the term jackknifing in the present context. Simulnet offers an automated jackknifing option with one of the network functions, the *Probabilistic Network*.

Training and Testing Exemplars

The first step in creating training and testing sets is to design the network exemplars, the examples of the relationship that the network is to learn. As mentioned, each exemplar is composed of predictor and criterion sections. The predictor section consists of a set of numbers each of which represent one predictor, or independent, variable, while the criterion section consists of a set of numbers each of which represents one criterion, or dependent, variable.

The predictor and criterion sections of each exemplar can be referred to as the predictor vector and the criterion vector respectively. The predictor vector describes the status of a cluster of independent variables. The criterion vector describes the status of the group of corresponding dependent variables. As a numerical example, suppose a network exemplar is designed as follows:

- The exemplar will contain three independent variables: amount of rain on a given day, mean temperature on that day, and mean barometric pressure on that day.
- The exemplar will contain two dependent variables: probability of rain on the following day and temperature the following day.

Measurements carried out on 2 particular days yielded values for the 3 independent variables of 0.0, 23, and 31, and values for the 2 dependent variables of 2.5 and 22. The predictor vector would simply be the set of numbers [0.0, 23, 31]. In the same way, the criterion vector would simply be the set of numbers [2.5, 22]. To extend the vector notation to the whole exemplar, the exemplar vector could be expressed as [0.0, 23, 31, 2.5, 22]. As a practical matter, when such an exemplar vector is presented to a network, the network will need to be told how many of the components of the vector represent predictor variables and how many represent criterion variables. Typically, a network exemplar comprised the values of p predictor variables and c criterion variables has the following structure:

$$[1, ..., p, p + 1, ..., p + c]$$

where elements 1 through p represent the values of the independent (predictor) variables, and elements p + 1 through p + c represent the values of the dependent (criterion) variables.

In this numerical example, it happens that all of the variables can vary continuously in value. Such variables are referred to as continuous variables. Other types of variables can also be represented within a neural network exemplar. These include discrete variables (for example, number of passengers), and categorical variables (for example, eye-color). Variables of any of these types can be combined within a single network exemplar. However, since ultimately the network can deal only with numerical information, categorical variables must first be converted into numbers by assigning them an arbitrary code. Such codes are referred to as dummy codes, or dummy variables. For example, if the categorical variables were eye-color, blue might be represented by one, black by two, and brown by three. There is, however, a danger with this way of representing categorical information: the network may tend to assign significance to the variables based on their numerical values. Thus, the network may tend to learn that there is a larger difference between blue and brown than between blue and black. Such value-based coding for categorical data is in general not ideal. A much more effective approach is to assign a single, unique output node to each category. Thus, in the present example, the network would contain three output nodes: One node representing blue, one representing black, and one representing brown. Correspondingly, our exemplars would each need to contain three dummy variables, one for each category. An exemplar whose criterion value represents blue would be coded as (showing only the criterion values) [..., 1, 0, 0]. An exemplar whose criterion value represents black would be coded as [..., 0, 1, 0], and one representing brown would be coded as [..., 0, 0, 1]. Straightforwardly, the first dummy variable represents blue, with zero denoting absence of that color and one denoting its presence. In the same way the second dummy variable represents black, and the third brown.

Within a single network data matrix, all of the exemplars must have the same format in terms of the number of independent and dependent variables. In other words, all of the exemplars in a training matrix, and in the associated testing matrix, must have the same number of predictor and criterion variables.

Memory considerations limit the maximum total number of predictor and outcome variables within any one exemplar. This generally does not present a real restriction. As mentioned earlier, if the classes in the training data are to be separable by more complex, nonlinear, decision boundaries, the network must be provided with more training exemplars then the number of network inputs. A network with, on the order of, several hundred inputs will need to be trained with at least several hundred exemplars.

Another consideration is that in order to achieve a generalization error of 10% for example, the network should be trained on up to 10 times as many training exemplars as there are network weights. As an example, with 100 input nodes and 2 hidden nodes, a network contains approximately 200 network weights. A generalization error of, say, 10%, would correspondingly require a training set consisting of 2,000 exemplars. With more hidden nodes, proportionately more training exemplars would be needed.

Creating Network Training and Testing Matrices

Simulnet provides a function to streamline the process of creating a network training or testing matrix. In order to make use of this function, it is only necessary that the predictor portions of each exemplar be defined. In particular, the predictor portion of each network training exemplar must be contained within a separate disk file. These files do not need to contain the criterion values for the exemplar. Criterion values can be added at the time that the network matrix is created, or at a later time. In order to use the Simulnet function that creates network training and testing files, two conditions have to be met:

- The values of the predictor variables for each single network exemplar must be contained within a single disk file. This file may be stored in ASCII (or text) format, or it may be stored in Simulnet's binary format.
- All of the files that are to be used in creating the network training or testing file must be located on the same disk drive, and in the same directory.

On the assumption that these two conditions have been met, the following procedure can be used to create a network training or testing matrix. This procedure will create a network training matrix using a set of 20 exemplar files that have been provided. These exemplar files represent EEG recordings that were made during an experiment in visual perception. The recordings were made in two conditions of the experiment. Consequently, the goal of creating a network file using these 20 exemplar files would be to see if a network would be able to correctly classify the files into two groups, corresponding to the conditions in which the files were recorded. The goal then is to see if a network can distinguish between these two groups of files. If a network can significantly well make this discrimination, there would be support for the notion that the experimental manipulation resulted in differences in the EEG signals recorded in the two conditions.

Procedure

1. Close all forms on the desktop. Select the *Create Training File* option from the *Network* menu. On the *Create Training File* dialog form click the *Example File: Select* button. This will show the *Select Files and Paths* dialog form from which exemplar files can be selected.

2. Select the exemplar files by entering the appropriate drive, directory, and file name settings in the *Source* group. The exemplar files should be located in a subdirectory, *\Data*, of the directory in which Simulnet was installed. By default, this subdirectory is *\Simulnet\Data*. To select multiple files, hold the *Ctrl* key down, while clicking on the file names that are to be selected. Select files *eeg01.dat* through *eeg20.dat*.

3. Using the drive and directory settings in the *Destination* group, select the path into which the training file is to be placed. In this exercise the completed training file will be placed in the *\Data* subdirectory of the directory in which Simulnet was installed.

4. Next, enter a value of 1 in the *Criterion: No. of columns* field. This will specify that each exemplar is to contain a single dependent, or criterion, variable.

   Enter a value in this field that is equal to the number of criterion variables that are contained in each exemplar. With a classification task, the number of criterion variables is equal to the number of categories being discriminated. An exception is the situation where there are only two categories. In that case only a single criterion variable is needed, dummy-coded 0 if the exemplar belongs to the first category, or one if the exemplar belongs to the second category. This is the situation in the present example.

5. Now, click the *Next* button. After a few seconds the *Criterion Values* dialog form will appear on the desktop. This form contains a matrix that has one column for each criterion variable, and one row for each exemplar file that was selected. The names of the exemplar files are shown on this matrix form as well. For each exemplar file that is in each row of this matrix, enter the criterion values for that exemplar. These values can be entered manually from the keyboard, or they may be pasted in from another matrix using the Simulnet *Copy* and *Paste* functions. In this procedure, dummy codes will be entered that will label the 20 exemplars as to whether they belong to condition 1 or 2. For files *eeg01.dat* through *eeg10.dat*, enter a dummy code of 0. For files *eeg11.dat* through *eeg20.dat*, enter a dummy code of 1. This dummy coding scheme should be used whenever the network is being asked to discriminate between two categories: The first category is dummy-coded with a 0 while the second category is dummy-coded with a 1. These dummy codes are listed for each file in the following table:

Network Classification Scores

File	Condition	Dummy Code
EEG01.DAT	1	0
EEG02.DAT	1	0
EEG03.DAT	1	0
EEG04.DAT	1	0
EEG05.DAT	1	0
EEG06.DAT	1	0
EEG07.DAT	1	0
EEG08.DAT	1	0
EEG09.DAT	1	0
EEG10.DAT	1	0
EEG11.DAT	2	1
EEG12.DAT	2	1
EEG13.DAT	2	1
EEG14.DAT	2	1
EEG15.DAT	2	1
EEG16.DAT	2	1
EEG17.DAT	2	1
EEG18.DAT	2	1
EEG19.DAT	2	1
EEG20.DAT	2	1

6. When the dummy codes have been entered, click the *OK* button on the *Criterion Values* dialog form. The training file will now be created. This process may take anywhere from several seconds to several minutes to complete, depending on the size of the file and the speed of the computer being used. When the training file has been created, a *Save File* dialog form will appear. On the *Save File* dialog form, choose the path and name under which the network file is to be saved. Then, press the *OK* button to save the network matrix.

To return to the previous step of this procedure, press the *Cancel* button on the *Save File* dialog form. Doing so will remove this dialog form and will reactivate the *Criterion Values* matrix form. Any required changes can then be made on the *Criterion Values* form. When these changes have been completed, press the *OK* button to create the network matrix. Note that the network matrix will not appear on the desktop during this procedure. However, once the network matrix has been saved as a disk file, the disk file can be opened as a matrix and the network matrix can then be examined.

7. A copy of the training file created in this example has been provided as file *eeg.trn*. To verify that the new training matrix has been constructed properly it can be compared with the matrix in file *eeg.trn*. To do this, open file *eeg.trn*, the provided network file, and the newly created network file onto the desktop. Then restore both matrices to normal size so that the contents can be examined, and position them so that they do not completely overlap. By switching from one matrix to the other, the values in the two matrices can be compared. In particular, compare the values in the last column of both matrices, column 2049, to check that these criterion values have been entered correctly in the new training matrix. File *eeg.trn* will be used later in this text in a neural network training example.

Neural Network Application: Predicting Earthquakes

The following example describes a hypothetical application for a neural network: How a neural network could be used to predict the value of a dependent variable, probability of earthquake activity, given exemplars from a historical record of the relationship between that variable and a number of independent variables, ground electric potential measurements. This example is only meant to illustrate the steps involved in using a neural network for prediction. It is not meant to represent an actual application.

Recordings are made, at the same time on each of 100 days, of measurements of the electric potential at 10 points over some geographic area. Each potential measurement represents a single independent, or predictor variable. One set of 10 such measurements represents the 10 predictor variables in 1 network exemplar. These predictor variables are continuous variables. To these predictor variables one criterion or outcome variable is added, representing the incidence of earthquake activity in the same geographic area on the day *following* the day of measurement. Incidence of earthquake activity will in this example be a categorical variable in that an earthquake either did or did not occur. This categorical variable is dummy-coded, by assigning the code 0 if no earthquake activity greater than some predefined level occurred, and 1 if earthquake activity greater than that level did occur. An alternative to treating earthquake activity as a categorical variable would be to use the strength of seismic activity in the area as a continuous variable. In any event, the entire exemplar will consist of 11 values; 10 predictor variables representing the electrical measurements made on 1 day, and 1 criterion, or outcome, variable representing the incidence of seismic activity on the day following the day of measurement. A typical exemplar constructed in this way would have the following form:

$$[1.23, 5.01, 0.12, 1.18, 0.44, 8.13, 2.00, 4.11, 9.10, 3.86, 0\,]$$

where the numbers 1.23 through 3.86 represent the potential measurements, and the final value of 0 represents the fact that earthquake activity in the area was below the predefined level.

Since recordings have been made on each of 100 days, the network training matrix will consist of 100 such exemplars. These exemplars are next separated into two groups. One group, the training set, will be used for training the network. The other group, the testing set, will be used for validating the network periodically throughout the training process. The testing set will be created by drawing at random, and without replacement, a total of 30 exemplars from the entire set of 100. The remaining 70 exemplars will then form the training set. The 70 exemplars chosen for training are next combined to form the training matrix, with each exemplar containing the 10 independent (predictor) variables in matrix columns 1 through 10, and the single dependent (criterion) variable in column 11.

$$
\begin{array}{c}
\\
\textit{exemplar 1} \\
\cdot \\
\cdot \\
\cdot \\
\textit{exemplar 70}
\end{array}
\begin{bmatrix}
1 & 2 & 3 & 4 & 5 & 6 & 7 & 8 & 9 & 10 & 11 \\
1.23 & 5.01 & 0.12 & 1.18 & 0.44 & 8.13 & 2.00 & 4.11 & 9.10 & 3.86 & 0 \\
 & & & & & \cdot & & & & & \\
 & & & & & \cdot & & & & & \\
 & & & & & \cdot & & & & & \\
5.12 & 0.01 & 4.42 & 9.12 & 7.10 & 3.08 & 6.82 & 2.55 & 6.20 & 1.13 & 1
\end{bmatrix}
$$

The 30 exemplars selected for validation are combined in the same way to form the testing matrix. These two matrices would then be saved as disk files.

The neural network is trained by repeatedly presenting it with the set of training exemplars. After each such training iteration, the values of network weights are updated. While the network is being trained in this way, it is validated by presenting it with the testing exemplars. Validation is generally carried out on some schedule, such as, for example, one validation iteration every five training iterations. Network error during validation is observed. When this validation error is seen to have reached a minimum value and to be starting to increase, training is stopped. The neural network can now be expected to be optimally trained, and to be ready to make predictions about novel exemplars for which no outcome information is available.

In order to run the network, that is, have the network make predictions for entirely novel exemplars, the network is supplied with exemplars that have the same format as that of the training and testing exemplars, but with the criterion variable left blank (in practice, set to zero). This of course is reasonable, since no criterion information will be available for such exemplars. The value of the outcome variables is in any case ignored when the network is being validated, or when the network is actually being run to make predictions. In the present example, such novel exemplars would be created, as before, by recording the electric field measurements in a single day. For each of these novel exemplars, the neural network's prediction is noted. In this example, this prediction represents the neural network's predicted incidence of earthquake activity on the day following the day of measurement. If it were indeed possible to train the network in this way, the network will then have achieved the ability to make short-range earthquake predictions.

Neural Networks as Data Analytic Techniques

The usefulness of neural networks in data analysis stems from their ability to operate as universal function approximators. As described earlier, a neural network containing at least a single hidden level with enough nonlinear nodes, can learn virtually any relationship between a set of independent and dependent variables that can be approximated in terms of a continuous function. With more than one hidden layer, relationships involving discontinuous functions may be approximated. Such networks thus have the power to model any functional relationship between sets of independent and dependent variables.

The capability of neural networks to operate as universal function approximators in turn derives from their ability to function as nonlinear analyzers. This ability is in part the result of the nonlinear activation function generally adopted for the hidden level nodes, such as the logistic function. Without some type of nonlinear activation function, neural networks would be reduced to performing only as linear analyzers, severely limiting their capabilities.

Consider the behavior of a network with only an input and an output level of nodes, and with a linear activation function for each node. With such a linear activation function the output, the activation value, of any node is linearly proportional to the sum of the inputs to the node. A network comprised of such linear nodes would only be able to present to the output nodes linear combinations of the values present at the input nodes.

Now suppose a third hidden level is added to this network, a level containing nodes that also have linear activation functions. The input to each of these hidden nodes will be a linear combination of the values applied to the input nodes. The output of each of these linear hidden nodes is then a linear combination of the input values, albeit in general scaled by some numeric factor. Each network output node, receiving scaled values from these hidden nodes, thus receives a set of linear combinations of network inputs, one such linear combination from each of the hidden nodes. Since the network output nodes also have linear activation functions, each output node linearly combines the linear combinations that it receives. This 'linear combination of linear combinations' of the values applied to the input nodes is, by the definition of linearity, simply another linear combination: The values generated by the output nodes are thus linear combinations of the original network inputs. No matter how many hidden layers a neural network had, if all nodes were linear the overall network could do no more than generate outputs that were linear functions of the inputs.

When the hidden-level nodes are given nonlinear, rather than linear, activation functions, the outputs of the hidden node now become nonlinear functions of the values applied to the input nodes. A consequence of this fact is that the way that any one hidden node responds to the signal from any one input node now depends on the signals coming to that hidden node from all other input nodes. In other words, the nonlinear activation function of the hidden nodes implies that interactions between the effects of the values applied to the input nodes will now occur. It is the complexity inherent in these interactions that is ultimately responsible for the rich, and sometimes chaotic, behavior that neural networks have been shown to have. In terms of their performance as classifiers, neural networks owe to this nonlinear activation function that ability to discriminate between classes of input vectors that are separated by nonlinear decision surfaces. A neural network can discriminate between classes that are separated by a nonlinear decision boundary in the variable space of the inputs because the network can form a correspondingly nonlinear decision boundary. The network can form such nonlinear decision boundaries because it can generate at its outputs nonlinear combinations of the data presented to its inputs. Experience has shown that real-world data sets are most often characterized by such nonlinear relationships among the features within the data.

Neural networks possess a number of advantages over competing data analytic approaches in the exploration of the functional relationships within a body of data. One advantage is that a neural network-based analysis is carried out without the constraints of an a priori model as the basis upon which the analysis is to be performed. A neural network can make use of, in principle, an unlimited number of features of the data being analyzed.

In contrast, any analytic technique that is constrained by a model will be limited to exploring only those aspects of the data that are addressed by the model. With Fourier analysis for example, the a priori model is that the sought-after relationships among the independent and dependent variables can be represented in terms of the frequency and phase components within the data. Similarly, with classification techniques based on cross-correlation or coherence, the initial assumption is made that the functional relationships within the data can be well represented by correlations or coherences. With neural networks there need not exist any preconceived notion about what aspects of the data are important for doing the classification. The data are presented to the network, and the network takes on the problem of determining what dimensions or features in the data hold the key to discriminability. The network's internal, distributed representation of the data, coded in terms of the network weights, contains the discrimination criteria.

This feature of neural networks as data analyzers can be a two-edged sword. While the internal representation of the training data can be more effective than preconceived criteria in terms of modeling the relationships within the data, these discrimination criteria developed by the network may not be easily accessed. That is, it may not be immediately obvious how the network's weights or hidden node activation values are related to physical features in the data. One reason that network weights are hard to interpret is that there is no unique set of weights for a particular network input-output mapping (Bioch, J. C., Verbeke, W. and van Dijk, M. W., 1994). Approaches to this problem of determining what it is that the network has learned include the use of sensitivity matrices. Sensitivity matrices are an attempt to quantify the extent to which each network input affects each of the network outputs (for example, Lisboa et al., 1994). Sensitivity matrices are discussed in the next section.

The extent to which the neural network can discover patterns within the data depends on a number of factors. A factor of primary importance is the number of exemplars that are available to the relationship to be modeled. In general, the likelihood that the network will be successful in capturing relationships within the data increases with the number of available exemplars. A second factor is the quality of the exemplars; that is, the extent to which the exemplars are contaminated by noise. The term noise in this context refers to features in the data that appear to be unrelated to the sought-after relationship. Included, of course, would be purely stochastic components to the data. An example of such a stochastic component would be noise of thermal origin at the output of an amplifier. A third factor is the nature of the data itself. How likely is it that the error surface defined by the data and the network configuration contains features such as local minima? While, as mentioned previously, local minima appear to be relatively rare in real-world data, the potential for these features to capture the network, and thus prevent it from training adequately, is nonetheless a possibility. A perhaps more real threat to the network's achieving its potential is an error surface with extended flat regions. Such regions might, in some cases, slow the rate at which the network was able to train to the point where the network is effectively stalled. As a practical observation, when a network does appear to have been trapped within one of these flat planes, the indication will be a relatively constant value of training error. In such a situation an appropriate strategy is usually to restart the network by reinitializing network weights and beginning training all over again.

A second advantage to using neural networks as data analysis techniques is computational efficiency in analyzing novel data. On the one hand, network training can take a substantial amount of time to complete. Once the neural network has been trained however, only a relatively small amount of computation is subsequently required to analyze a novel exemplar. This advantage would be particularly important in real-time applications where the network could, for example, rapidly classify signals into one of several categories.

Analyzing the Network Weights

An important question in many network applications is: What is it that the network has learned? That is, what is the basis for the performance of the network in some particular application? One approach to try to answer this question is to compute what is referred to as a sensitivity matrix. A sensitivity matrix contains a description of how sensitive each of the network outputs is to changes in each of the network inputs. Thus, for each network output, the sensitivity matrix tells us how much the value of that output would vary as each of the network inputs was varied. In this way, we obtain a characteristic pattern, or signature, for that output that reflects the contribution to that output of the value of each network input. In Simulnet, the *BackProp Network* and *Genetic Network* functions provide the option to generate such a sensitivity matrix. The sensitivity matrix has one column for each network output, and one row for each network input. The value of an element at row r and column c is a relative measure of how much output c is affected by input r. These values are relative in that they are a dimensionless index that can be used to compare, for each network output, the relative effect of each input.

Sensitivity matrices in Simulnet are computed by evaluating the matrix of partial derivatives of matrix outputs with respect to network inputs, known as a Jacobian matrix (Lisboa et al., 1994),

$$J_{ik} = \partial y_k / \partial x_i$$

where J_{ik} is the element of the sensitivity matrix that describes how sensitive output k is to input i; x_i is the i-th network input; y_k is the k-th network output; and ∂ denotes partial differentiation of a function, in this case y_k, with respect to one of the variables of that function, in this case x_i.

The following illustration will attempt to explain the meaning of the partial differentiation of a function with respect to one of its variables. Picture the value of a network output y_k (the function that is being differentiated) as the height of a surface, which we can term the output surface, in a space defined by a number of dimensions, each of which corresponds to one of the network inputs, x_i (the variables of the function). This is a similar concept to that of the error surfaces that were discussed earlier. In the present case, the output surface will also be characterized by some topography. In particular, at any point on the output surface, the output surface will have a value of slope along each of its dimensions. If there are two dimensions, there will be two values of slope. The slope along any one dimension is the partial derivative of the function y_k with respect to that dimension. This is the meaning of the term partial derivative. The partial derivative of a multivariate function, a function of more than one variable, with respect to one of its variables is the slope of the function along the dimension corresponding to that variable.

To actually compute the values in the sensitivity matrix, mathematical techniques can be applied to expand and simplify the preceding definition of a Jacobian matrix. The result is given by the following expression:

$$S_{ik} = \Sigma_j \, w_{ij} \, (1 - h_j) \, h_j \, w_{jk}$$

where S_{ik} is the element of the sensitivity matrix corresponding to the i-th input and the k-th output, w_{ij} is an input-to-hidden weight, w_{jk} is a hidden-to-output weight, and h_j is the output value of the j-th hidden node.

A convenient way to examine the contents of a sensitivity matrix is to graph the matrix using the *XY Graph* function in Simulnet. On this graph, each graph line will correspond to a single matrix column, and thus a single network output. The graph line for column 1 will correspond to network output 1, and so on. The x axis of the graph will represent the number of the matrix row, and thus the network input number. Here, point 1 on the x axis will represent row 1, and hence network input 1, and so on. The height of a graph line for a particular network output, at some x axis value, is an indication of how sensitive that output is to the network input corresponding to that row: The higher the graph line at a particular point on the x axis, the more sensitive is that network output to changes in that network input.

Summary of Neural Network Operation

To summarize the general operating principles of neural networks as a data analytic approach: in a typical application, a neural network configuration typically consists of three layers of nodes. An input level containing a number of nodes equal to the number of elements in a predictor vector simply provides a set of connection points allowing the values in a predictor vector to be supplied to the network. Variable-strength couplings, the network weights, connect the input level with the second level of nodes, the hidden level. These input-to-hidden layer weights are modified over the course of the network training phase using some rule, typically some form of the error backpropagation algorithm. After some amount of training, the hidden-level nodes will come to represent features abstracted by the network from the bolus information within the training exemplars. A second set of weights, also modified during the training process, connects the hidden level with the output level of nodes. The nodes in the output level thus each receive a unique weighted combination of the internal features stored in the hidden level. In turn, the output nodes typically generate a value that represents the weighted sum of these weighted combinations.

In the training phase, the neural network is trained by presenting it with a series of exemplars. Each exemplar consists of two parts: a predictor vector representing the values of a group of independent variables, and a criterion vector representing the values of a set of associated dependent variables. As an example, the independent variables might be recordings of brain electrical activity. An associated dependent variable could then be a code representing a corresponding behavioral response, or experimental condition such as a cognitive or perceptual task. Using a training rule such as error backpropagation, the neural network attempts to minimize the difference between the actual outcome, the output, of the network and the target outcomes that are coded within each criterion vector section of the training exemplars. To accomplish this goal, the network develops an internal representation, over the course of the training session, of the features present in the training exemplars. As mentioned earlier, one limitation that can also be an advantage in applying neural networks to such pattern analysis tasks is that these internal feature representations may not correspond to features of the predictor vectors in any obvious way. The positive side of this behavior is that these internal representations may, given sufficient training, come to represent features of the predictor vectors which are more efficient in performing the pattern analysis task than other features that might have been presumed to be significant on the basis of beforehand assumptions. In any event, these internal representations are distributed in the network weights, and are summarized in terms of the activation values, the outputs, of the hidden nodes. For each hidden node, these activation values or outputs are the weighted sums of the outputs of the previous level, the weightings being the network weight values.

As training proceeds, the learning progress of the network is periodically validated by presenting the network with a set of test exemplars, while recording the resulting test errors. The set of test exemplars is generally created by sampling at random and without replacement from the total pool of training exemplars. By choosing test exemplars in this way there is some assurance that the test exemplars do not significantly differ from the training exemplars. For each test exemplar, test error is recorded as the difference between the actual network outcomes and the target outcomes coded in the test exemplars. The magnitude of the test error indicates how well the network has abstracted the significant features in the training exemplars or, in other words, how well the network has learned.

When test error reaches a minimum, the network can be considered to have been optimally trained. At this point the network can be operated in the running phase. In the running phase the network is put to work by presenting it with novel exemplars for which there are no known criterion values, or outcomes. The network will then generate an output for each of these novel exemplars on the basis of the information that the network has abstracted from the training exemplars, over the course of the training. These outputs are the network's predicted outcomes for each of the novel exemplars.

Case Studies: Neural Networks in Neuroelectric Signal Analysis

This section will review a selection of studies that illustrate the applicability of neural networks to the analysis of experimental data. The studies reviewed here demonstrate the effectiveness of neural network systems of several different types in the classification and categorization of EEG data.

Gabor and Seyal (1992) applied a multilayer backpropagation neural network to the task of learning to recognize a particular pattern of voltage changes in the EEG of epileptic patients. A typical finding with epileptic patients is that their EEG measurements show a quite characteristic voltage pattern during times when the patients are not suffering a seizure. On a graph this voltage pattern appears as a sharp spike, followed by a slower more rolling wave-like feature, and is therefore referred to as a spike-wave pattern. For each of five epileptic subjects, the authors recorded EEG's from eight pairs of channels that included all scalp regions. The neural network used for the analysis consisted of eight input nodes, one for each of the eight EEG channel pairs, and a single output node. The eight input nodes each received the signal from one of the eight EEG channel pairs. The single output node represented the single categorical dependent variable in this study, incidence of the spike-wave pattern. The network included a hidden level with eight nodes. To decrease the computational load on the network, EEG data was preprocessed by calculating and using only the slopes of the spike events for each of the eight channels. The training and testing vectors corresponded therefore to the spatial distribution of the rates of change of the spike-wave event. Such preprocessing is typically carried out on raw data before the data is presented to the network. The general idea is that if there is reason to believe that some of the features in the data are more relevant to the network learning task than others, the network can be trained more efficiently and quickly using only this subset of the information in the data. In this EEG study, the authors found that the network was able to correctly classify an average of 94% of the waves, with only 21% false-positive classifications.

In a study with a similar goal, Jando, Siegel, Horvath and Buzaki (1993) also used a multilayer network to classify EEG signals into two categories on the basis of whether or not they contained epileptiform activity. Spike-wave activity was recorded over 12 hours from the neocortex of rats. The design of the neural network was optimized by conducting a parametric study in which the number of input and hidden neurons was varied. In a parametric study the behavior of a system is assessed as the values of one or more system parameters are varied over some range. The authors found that a network of 16 input and 19 hidden nodes had the best classification performance. One output neuron was used, representing occurrence of the spike-wave activity. The authors analyzed both the raw EEG time series, as well as Fourier transformations of the raw data. Each time series consisted of 12 seconds of recorded EEG data, digitized at 100 samples per second, and selected visually to represent 1 of the 2 data conditions. The training set consisted of 469 time series containing spike-wave activity, and 1,133 time series not containing spike-wave activity. The authors chose to analyze the time series using a moving time-window 16 data points wide, sliding across the time series in steps of 1 data point. At each step, the 16 data points within the window were presented to the network input nodes. This strategy allowed the authors to see how well the network could learn to find the spike-wave pattern as a function of time along the record.

The network was able to correctly classify 96% of epileptiform events, while misclassifying 30% of nonepileptiform events. The authors suggest that this performance demonstrates the power of a nonlinear analytic technique, such as a neural network, to find correspondingly nonlinear relationships between dependent and independent variables. Techniques limited to utilizing only linear relationships are prone, they say, to committing false-positive misclassification.

The researchers found also that the network trained more quickly with the data that had been preprocessed by a Fourier transformation than with the raw data. This would suggest that the distinguishing features between epileptiform and control waveforms consisted of amplitude and phase differences of periodic components within the data. As Smith (1993) points out, doing some of the work that the network would otherwise have to do by preprocessing the data usually results in improved training. Generally, however, preprocessing data before presenting it to the network will only improve network training performance if the preprocessing is able to help uncover features in the data that are genuinely relevant to the network learning task. The right sort of preprocessing will extract features from the data that will help the network to train. In other words, preprocessing the data to help the network to train should only be done on the basis of some hypothesis about what features in the data are important to the relationship that the network is to learn. If the hypothesis is incorrect, and the wrong sort of preprocessing is done, the network can be prevented from training at all. It should be emphasized that if there is no independent evidence for knowing what sorts of features within the data are important for what the network is to learn, then no attempt should be made to preprocess the data.

In a study of EEG recorded during sleep, Grozinger, Kloppel and Roschke (1993) used a multilayer network to classify EEG waveforms according to whether they showed evidence of REM (rapid-eye-movement) sleep. EEG recordings were made from a sleeping subject from the vertex (at the top center of the scalp), digitized at 100 samples per second, and separated into 6 frequency ranges; 0.5 to 3.5 Hz, 3.5 to 7.5 Hz, 7.5 to 15 Hz, 15 to 25 Hz, 25 to 45 Hz, and 0.5 to 45 Hz. The raw EEG data was preprocessed by computing power level within each band. The network then consisted of six input nodes, one for the power level of each frequency component, four hidden nodes and one output node. Each exemplar consisted of 2,048 data points. Data from 1 night was used for training and data from a subsequent night used for testing, with a total of 1,300 exemplars. The network correctly classified an average of 89% of test exemplars. The authors point out that conventional analysis to determine sleep stage requires additional recordings of electrooculographic and electromyographic potentials. The neural network allows such classification to be carried out using EEG recordings alone.

Grozinger, Kloppel and Roschke (1993) also used a multilayer network to distinguish patients with multiple sclerosis (MS) from controls. The basis of the classification was an EEG feature known as the P300 response. The P300 response is a positive waveform (P) occurring about 300 milliseconds (300) after a stimulus is presented. Characteristically, the amplitude of the P300 has been shown to be sensitive to stimulus probability, increasing for rare stimuli, while latency (the time between stimulus and response onsets) has been found to be linked to task difficulty. P300 latency increases, for example, with increasing difficulty of stimulus discrimination. P300 characteristics have been found to be altered in patients with MS with, in particular, an increase in latency. Such alterations have, however, generally been too subtle to allow them to be used in clinical diagnosis. In this particular study, an 'oddball' paradigm was used to elicit the P300: Infrequent, or oddball, target audio tones were interspersed with frequent standard tones of a different pitch. Recordings of P300 components to the target tones were made at three electrode sites along the midline of the scalp; Fz, Cz, and Pz. Averages were then formed over 100 recordings, with each recording containing 256 data points. Network training and testing exemplars were then constructed by uniformly sampling 25 points from each of these averaged recordings. A set of 101 training exemplars was formed in this way for each of the 3 electrodes, with 51 exemplars from the MS group and 50 from the control group. Testing data consisted of 10 MS and 10 control exemplars for each electrode. The authors chose to use separate, but identical, neural networks for the data from each of the three electrodes. Each network consisted of 25 input nodes (corresponding to the 25 exemplar data points), 8 hidden nodes, and 2 output nodes. Final scoring was done on the basis of a two-out-of-three majority rule. An exemplar was categorized according to whatever classification was assigned to that exemplar by at least two of the three networks. Classification accuracy, when the performance of each network was considered separately, was found to be 85% at Cz, and 80% at both Fz and Pz. Using the majority rule, classification accuracy was 90%. The authors admit that a difficulty with using neural networks is that the basis of the classifications is generally not easily accessible. On the other hand, if the networks are able to effectively distinguish between disease and control conditions, they can nevertheless be useful diagnostic tools.

The diagnostic capability of neural networks has also been used to classify subjects into one of three clinical categories; depressive, psychotic, or normal (Kloppel, 1994b). EEG recordings were made from 18 subjects: 6 depressives, 6 psychotics, and 6 normal controls; from 16 scalp electrodes, over an interval of 30 minutes. Preprocessing consisted of reducing 4 second segments of the EEG record to 6 values, representing spectral power levels in 6 frequency bands, ranging from 0.5 Hz to about 30 Hz. The neural network was examined after training on data from varying numbers of subjects. After training on data from only 1 subject, the network was able to classify unlabelled data segments from that same subject with an accuracy of 80%. After training on two or more subjects however, classification accuracy for data from any 1 subject dropped to 66%. The network was able to recognize data belonging to the subject on which it had been trained, but was only marginally able to generalize this knowledge to the classification of data from other subjects.

A different type of network, a Learning Vector Quantizer (LVQ) developed by Teuvo Kohonen (1989), has also been used in studies involving EEG classification. The LVQ is a network composed of mutually interconnected nodes. Multilayer backpropagation networks, in contrast, are organized in hierarchically connected layers. As for error backpropagation networks, a set of training exemplar vectors is presented to the LVQ. After a number of such presentations the LVQ is then able to place a novel exemplar vector into one of several categories. This classification function is similar to that performed by traditional statistical cluster-analysis techniques.

Pfurtscheller, Flotzinger, Mohl and Peltoranta (1992) used an LVQ to predict laterality of hand movement. Thirty channels of EEG data were recorded from three subjects prior to voluntary right or left hand movements. Subjects were asked to press a microswitch with either a left or right finger after a cue. EEG's were recorded in the interval between cue onset and the start of the movement, in the 8-to-10 Hz and 10-to-12 Hz frequency bands. In a training phase, the LVQ was allowed to self-organize, a process analogous to the pattern formation that occurs in multilayer neural network weights. After training, the LVQ could significantly well predict side of hand movement, with an accuracy of 85%, 74%, and 64% respectively for the 3 subjects. Significantly, the authors claim that this finding is the first demonstration that EEG signals can be classified without the use of averaging. They point out that by not using averaging the problem of dealing with a statistically nonstationary signal is avoided.

The LVQ network has a number of advantages in the classification of EEG (Flotzinger, Kalcher and Pfurtscheller, 1993). First, LVQ classification is fast in comparison with, for example, error backpropagation networks. This property allows classification to be carried out in real time. Second, LVQ training is similarly fast, permitting on-line learning to be carried out. That is, the network can accept training exemplars as fast as they are generated by the EEG recording equipment. Third, the LVQ network has fewer parameters whose values need to be set to achieve an optimal level of training. Typically, only a learning rate parameter, governing the degree to which network weights are updated on presentation of each training exemplar, needs to be adjusted.

Summary and Discussion of Case Studies

Neural networks, both of the multilayer and LVQ types, have been demonstrated to be effective as pattern classifiers. Using multilayer networks, this ability has been applied to the problem of recognizing epileptic spike-waves in humans (Gabor and Seyal, 1992) and in rats (Jando, Siegel, Horvath and Buzaki, 1993); classifying stages of sleep (Grozinger et al., 1993); recognizing the effects of multiple sclerosis on event-related potentials (Slater et al., 1994); and distinguishing between normal, depressive, and psychotic subjects (Kloppel, 1994b). Learning Vector Quantizer networks have been used to perform real-time classification of movement-related potentials in order to predict the laterality of finger movement (Pfurtscheller et al., 1992).

One reason that neural networks can effectively categorize time-series data may be that they function as nonlinear discriminant analyzers. Webb and Lowe (1990), for example, show that this discriminatory ability is a result of the first half of the network, from input nodes to hidden nodes, performing a nonlinear transformation of the input data into a feature space, defined by the hidden units, in which the discrimination between categories should be easier. The second half of the network, from the hidden to the output nodes, then executes a linear transformation aimed at minimizing the mean-square error to a set of given output patterns. In short, neural networks are capable of performing nonlinear discriminant analysis. The brain, as a distributed, nonlinear dynamical system, is probably not effectively describable or analyzable using purely linear methods. The analysis of complex nonlinear systems, such as the brain, can reasonably be expected to require the use of correspondingly nonlinear methods, examples of which have been discussed in this section. See Kloppel (1994a) for a comprehensive overview of the application of neural networks to the analysis of EEG data.

The Simulnet *BackProp Network* Function

In this section, the various controls and settings that are available with the *BackProp Network* function are listed and described. The *BackProp Network* function implements a supervised learning algorithm: A multilayer feedforward network trained using a form of gradient descent, the error backpropagation learning rule. Table 2.1 lists and describes the available controls and settings located on the *BackProp Network* and the *BackProp Network Options* dialog forms.

Before attempting to use the *BackProp Network* function, the reader is strongly encouraged to run Simulnet, select the *BackProp Network* function from the *Network* menu to show the *BackProp Network* dialog form, and then study the information in the following table. Going through this exercise is particularly important if the reader wishes to use this function with data other than that which has been supplied with Simulnet.

Table 2.1 *BackProp Network* Controls

Controls Located on *BackProp Network* dialog form

Control	Description
Training Data	
Training exemplars	The network can be trained on any contiguous set of exemplars (rows) in the training matrix. The first exemplar is specified by entering the appropriate value in the *Training Data: First Example* data entry field. The last exemplar is similarly specified using the *Training Data: Last Example* data entry field.
Training File Analysis	The network training file can be analyzed for two features: outliers and nonnormality. Outliers are defined as data values that are greater than four standard deviations away from the matrix mean. Normality is assessed by computing the skew over the matrix.
	When this function is invoked, nonnormality will produce a recommendation about a matrix transformation that can be used to correct the skew. The user will have the option of allowing Simulnet to carry out the indicated transformation; the value of skew following transformation is displayed.
	The number of outliers beyond four standard deviations from the mean is computed and displayed. The user has the option of allowing Simulnet to trim these outliers. When trimmed, outliers

above and below 0 are given values of +4 or -4 standard deviations respectively, from the mean.

This feature is accessed using the *Analyze training file* button.

Testing Data

Testing exemplars	The network can be tested on any contiguous set of exemplars (rows) in the testing matrix. The first exemplar is specified by entering the appropriate value in the *Testing Data: First Example* data entry field. The last exemplar is similarly specified using the *Testing Data: Last Example* data entry field.

Composition of Training and Testing Exemplars

Predictor values	Each exemplar is assumed to consist of a predictor portion, in the first set of columns of the training and testing matrices, followed by a criterion portion, in the last set of columns. For instance, an exemplar may consist of 10 columns; the first 9 columns being the predictor portion, and the final 1 column being the criterion portion. Criterion columns are always assumed to be at the end of the matrix columns. The number of criterion columns is specified using the *Examples: Criterion columns* field.
	The network can be trained and tested using only some range of (contiguous) columns within the predictor portion of an exemplar. Using the example above, the network could be trained and tested using only columns 3 to 7 of the predictor portion of each exemplar. This would not affect the criterion. The criterion portion would still be the final one column (in this example).
	The first predictor column to be used is specified by entering a value for the *Examples: Predictor: Start column* field. The last column is specified using the *Examples: Predictor: End column* field.

Network Architecture

Input and Output nodes	The number of input and output nodes is fixed by the specified number of criterion and predictor values: The number of input nodes is equal to the number of predictor values; the number of output nodes is equal to the number of criterion values. These values are displayed on the dialog form for the information of the user only. They can not be altered.
Hidden nodes	The number of hidden nodes may be set to any value from 1 to approximately 32,000. The maximum value will in practice be

limited by the size of available memory. A prominent question in setting up a neural network is how many hidden units to use. This value is set using the *Network Size: Hidden nodes* field.

Simulnet provides a function that will estimate the number of hidden layers on the basis of the number of significant eigenvalues in the training matrix. This function is invoked using the *Get Hidden Nodes* button. Simulnet provides an additional function to estimate not only the number of hidden nodes, but also the values of the input-to-hidden layer weights. This function is invoked using the *Get Nodes and Wts* button.

Hidden layers

The number of hidden layers can be specified as 1 to approximately 32,000. The actual number will be limited in practice by available memory. It has been shown that a single hidden level with enough hidden nodes in it can approximate any network input-output function. In some cases however, in particular with problems involving discontinuous functions, or boundaries, between classes, a network with more than one hidden level may be more effective than a single hidden-level network. The number of hidden layers is specified using the *Network Size: Hidden layer: no. of layers* field.

Network Analysis

Sensitivity matrix

A sensitivity matrix is a matrix that describes the extent to which each network output is influenced by each of the network inputs.

The sensitivity matrix has one column for each network output, and one row for each network input. The value in the sensitivity matrix at row r and column c is an indication of how sensitive network output c is to network input r. For example, if the neural network has 2 outputs and 10 inputs, the value in the sensitivity matrix at column 1 and row 3 indicates how sensitive network output 1 (column 1) is to network input 3 (row 3). The values in the sensitivity matrix are relative values: The higher the value in the matrix, the more sensitive is a network input to a network output.

Network Archiving

Saving a network

A network configuration can be saved to disk at any time using the *Save Net* button.

A network configuration consists of the values of the weights in all layers, along with a header specifying the network configuration: the number of input, hidden and output nodes, and the number of hidden layers.

Recreating a network	A network that has been saved as a disk file can later be recreated using the *Load Network: File* button. Using these functions, network training can be stopped at any point, and the network saved to disk. The network can then be recreated at any future time and training resumed. This feature allows network weights to be initialized from a file, rather than by means of a pseudorandom number generator.

Display Objects

Status	Selecting the *Display: Status* option shows the number of the current training iteration and the *RMS* and *AVErage* training and testing errors, updated every training and testing iteration. The *Correct* indication on this display represents the percentage of output nodes with achieved values that are within the respective criterion level of the respective target outcome values in the training and testing sets.
Error History	Selecting the *Display: Error History* option shows a graph of AVErage training errors (color is default plot color 1) and AVErage testing errors (color is default plot color 2) from the start of training. The number of iterations per graph is specified by the value of the *Passes* parameter. When this number of iterations has been graphed, the graph is cleared and restarted beginning with the next iteration.
Data Matrix	Selecting the *Display: Data Matrix* option shows a matrix containing target and predicted values for each of the criterion variables, in each exemplar, in both training and testing matrices. These values are arranged in four groups of columns, and with one row for each exemplar. If n is the number of criterion variables, columns 1 to n contain the target values from the training set. Columns n + 1 to 2n contain the corresponding predicted values. Columns 2n + 1 to 3n contain the target values from the testing set. Columns 3n + 1 to 4n contain the corresponding predicted values. For instance, if there are two criterion variables, the first two columns contain the target training values, and the next two contain training predicted training values. Columns 4 to 6 contain target testing values, and columns 6 to 8 contain predicted testing values.
Network	Selecting the *Display: Network* option displays a color-coded representation of the network structure. Nodes are represented by rectangles, and are joined by color-coded links representing network weights. A legend shows the range of weight values corresponding to the colors. Network input nodes are at the top of the display, outputs at the bottom.

Weights	Selecting the *Display: Weights* option displays a color-coded representation of weight values. Each block of weights is shown as a separate block on this display, beginning with the weights between the input nodes and the first hidden level. For this first block, the weights are arranged as a matrix with input nodes represented by columns and hidden nodes in the first hidden level represented by rows. If more than one hidden layer is present, a separate block is displayed for each of the blocks of weights between the hidden layers. For these blocks, weights are arranged as a matrix with hidden nodes of the level closest to the input represented by columns, and the nodes of the following hidden level represented by rows. There is always one block displayed showing the weights between the penultimate level and the output level. In this final block, weights are arranged as a matrix with the penultimate level nodes represented by columns and output nodes represented by rows.
Histogram	Selecting the *Display: Histogram* option displays the distribution of weight values. The horizontal axis represents weight values, while the vertical axis shows the corresponding number of weights in each weight value interval.

Training Schedule

Continuous vs. Single step	Network training can run continuously, or can be iterated in single steps. One training iteration consists of presenting the network with all of the training exemplars. To single-step training, select the *Displays: Single-step* option.

Controls Located on *BackProp Network Options* dialog form

Training Parameters

Learning Rate	Learning rate can be preset at a fixed value, or allowed to vary with a unique value for each node. Training error convergence with many types of training data should generally be faster with the variable, adaptable, learning rate option.
	To use fixed learning rate, select the *Training: Learning rate: Adaptive* option. The same value of learning rate is used for all nodes. This value is entered in the associated data entry field.
	To use variable learning rate, select the *Training: Learning rate: Adaptive* option. A unique learning rate value is computed for each node. This value is updated in each training iteration based on that node's on-going training performance. The initial value of learning rate is the value entered in the *Training: Learning rate: Set* data entry field.

Damping	This parameter is also known as momentum. Using damping can accelerate training error convergence for problems with error surfaces that contain steep ravines. Damping implements a time-averaging of the weights, with an actual weight-change consisting of a portion of the computed weight change (1—Damping) together with a portion of the previous weight change (Damping). To disable damping, set its value to zero. Damping is selected using the *Training: Damping* option.
Noise Perturbation	This option is an attempt to speed the convergence time for difficult problems. The rationale for this option is that a better value for a weight may be found by perturbing its value by an amount proportional to the degree to which the weight is contributing to the current training error. If the weight is contributing greatly to the current error, perturb the weight by a correspondingly large amount. The value of the perturbation is determined by a pseudo-random Gaussian noise source, and so the perturbation may push the weight even further from the best value. If the weight currently has a poor value however, little will be lost by making its value even further away from optimum. Training sets that take a long time to converge to a low value of error may benefit from this option.

The following two parameters govern the operation of this option.

Training: Noise level: Sets the initial level of the noise perturbation. This noise level is modified during training by a factor inversely proportional to the weight's contribution to node error. Thus, weights that contribute proportionately more to the error are given a larger noise perturbation than nodes that contribute less error.

Training: Noise decay: Sets the rate of exponential decay of the noise level; this value is ignored if noise level is set to zero.

Training/Testing Schedule

Testing rate	The testing rate is the frequency with which testing is carried out. For example, if testing rate is set to five, then testing is performed once every five training iterations. This value is set using the *Training: Test rate* field.

Training Termination Criteria

Training error	When the *Stopping Criterion: Train error* option is selected, training continues until training error reaches a specified value.

This specified value is the error value entered in the *Stopping Criterion: Train all correct* field. This is the error level below which an output node is judged to have achieved a correct response during training. A value of 0.1 means that to be judged as correct, an exemplar must generate on all output neurons an output value that is within 10%.

Testing error	When the *Stopping Criterion: Test error* option is selected, training continues until testing error reaches a specified value.
	This specified value is the error value entered in the *Stopping Criterion: Test all correct* field. This is the error level below which an output node is judged to have achieved a correct response during testing. A value of 0.1 means that to be judged as correct, an exemplar must generate on all output neurons an output value that is within 10%.
Iterations	When the *Stopping Criterion: Iterations* option is selected, training continues until the specified number of iterations has been completed.
Time	When the *Stopping Criterion: Time (mins)* option is selected, training continues until the specified number of minutes has elapsed.
All output nodes train correctly	When the *Stopping Criterion: Train all correct* option is selected, training continues until all output nodes are within the error value specified by the *Stopping Criterion: Train error* value
All output nodes test correctly	When the *Stopping Criterion: Test all correct* option is selected, testing continues until all output nodes are within the error value specified by the *Stopping Criterion: Test error* value

Network Node Parameters

Gain	The value of gain determines the shape of the node transfer function for the logistic, hyperbolic tangent, and linear truncated functions. Gain determines how sensitive the outputs of these functions are to changes in the inputs. Gain is specified using the *Network Nodes: Gain* field.
Bias	The value of bias is the amount that the weighted sum of the node's inputs must exceed for the node output to rise above zero. This value can generally be set to -1. Setting this value to zero will restrict the range of functions that the neural network can compute, thus restricting the ability of the network to train.
	If the network is considered as computing a polynomial approximation to the functional relationship distributed among the

network weights, then bias is analogous to the constant term in the polynomial. Bias is specified using the *Network Nodes: Bias* field.

Transfer (activation) Function

Nodes in the hidden and output layers can be given any one of four different transfer, or activation, functions. All the nodes in any one level will have the same function. Activation functions are selected using the *Hidden Node: Function name* option, or the *Output Node: Function name* options, where *Function name* is one of the following:

Function name	Equation; x is the net input to node; f(x) is node output
logistic	$f(x) = 2 / (1 + \exp(-gx)) - 1$ (where g = node gain)
hyp. tangent	$f(x) = \tanh(gx)$ (where g = node gain)
linear trunc.	$f(x) = gx; -1/g \leq x \leq 1/g$ $f(x) = -1; x < 1/g$ $f(x) = 1; x > 1/g$ (where g = node gain)
binary	$f(x) = 1; x \geq 0$ $f(x) = -1; x < 0$ (where g = node gain)

Weight Initialization

Distribution

Weights are initialized using pseudo-random noise values. This noise can have optionally a Gaussian or uniform distribution, selected using the following options:

Weights: Distribution: Gaussian: Weights are initialized by randomly sampling from a Gaussian distribution with mean of zero and standard deviation specified by the *Range* parameter.

Weights: Distribution: Uniform: Weights are initialized by randomly sampling from a uniform distribution with the range specified by the *Range* parameter.

Range

This value determines the range used for initializing weight values. If this parameter is set to zero then an estimate of the optimum value of range will be computed when weight initialization is carried out. Range is set using the *Weights: Range* field.

Reinitialization

Reinitialize network	This function, invoked using the *Restart* button, reinitializes the network using pseudo-random noise with characteristics set using the selected options for *Weight: Distribution* and *Weight: Range*.

Simulnet Exercise: Predicting Chaotic Data 1

This exercise will illustrate the procedure to be followed in order to use the Simulnet *BackProp Network* function. This function implements a multilayer neural network trained with the error backpropagation rule. The network will be trained to predict the value of a chaotic time series, given the previous value of the time series. That is, given the current value of a chaotic variable, the network will learn to predict the next value of the variable. The chaotic time series was generated using the logistic map. The logistic map is a function which is defined by the difference equation,

$$y_{n+1} = y_n + G\,(y_n\,(1 - y_n))$$

The behavior of this equation is determined by the control parameter G. When G is set to zero, the value of x does not change as the equation is iterated. Increasing the value of G results in a number of different behaviors. In particular, when G reaches about 2.57, the logistic map generates a large number of apparently random values. This behavior of the logistic map is termed chaotic. The data used in the present exercise were generated using a value of G = 3.0.

The following procedure was used to generate the network training and testing data used in this exercise. For each data set, the logistic equation was iterated 200 times using a random starting value, thus generating a chaotic time series of 200 data points. These 200 data points were then formed into 100 pairs of points: Each pair of contiguous data points in the original set of 200 created 1 pair in the new set of 100 pairs of points. These 100 pairs of points were organized as a matrix with 2 columns of 100 rows. In each row, column 1 contained the first data point of each pair, while column 2 contained the second data point of each pair. In terms of the original chaotic time series, within each row, the data points in column 2 immediately follow in time sequence the data points in column 1, as follows:

$$
\begin{array}{l}
\textit{exemplar 1} \\
\textit{exemplar 2} \\
\quad . \\
\quad . \\
\quad . \\
\textit{exemplar 100}
\end{array}
\left[
\begin{array}{cc}
\text{time-point 1} & \text{time-point 2} \\
\text{time-point 3} & \text{time-point 4} \\
. & . \\
. & . \\
. & . \\
\text{time-point 199} & \text{time-point 200}
\end{array}
\right]
$$

Each row of this matrix forms 1 network exemplar: For each exemplar there is 1 independent variable in column 1 and 1 dependent variable in column 2. The network will therefore be asked to predict the next value that would be generated by the logistic equation, given the current value being generated by the equation.

Two sets of 100 pairs of data points were generated using this procedure. Each set was constructed to be independent of the other by starting the logistic equation each time with a new and random initial value. The network will be trained on one of these sets, and tested, or validated, using the other set. As the network is being trained, validation, or testing, will be periodically carried out by applying the testing data once every five training iterations, or passes through the training data.

Procedure

1. Close all forms on the desktop. From the *Network* menu, select the *BackProp Network* option. On the *BackProp Network* dialog form click the *Load setup* button. On the *Load Setup* dialog form select file *lognn.set*. This file contains the setup values that are needed to run this exercise. In particular, a training file, *log.trn*, and a testing file, *log.tst*, will be loaded. Preset values for network parameters will be installed as well.

2. On the *BackProp Network* dialog form, click the *Train* button. A number of objects will appear on the desktop: a status form showing the current network training and testing errors and the pass number, a matrix form that will be used to hold the network results, and a graph form on which will be plotted a record of the network training and testing errors averaged over the corresponding exemplars. The network will then start training.

   As the network trains, training and testing error should eventually start to decrease, and training and testing correct percentages should rise. These figures are the percentages of the training and testing exemplars for which the network has achieved a score less than or equal to the criterion level. This criterion level is specified as 0.1 on the *Options* dialog form. In other words, an exemplar is judged to have been scored correctly by the network when the value that the network generates for that exemplar is within 10% of the target value for that exemplar as specified in the training and testing files. The number of passes needed to train the network to criterion cannot be determined precisely, but it will probably take between 100 and 200 passes.

   The graph showing the network errors should resemble Figure 2.9, Error History. This graph shows two plot lines: one for training error averaged over all training exemplars and one for testing error averaged over all training exemplars. The graph shows the network training to criterion in about 45 passes through the training exemplars. In practice, however, it may take up to 150 passes to train the network.

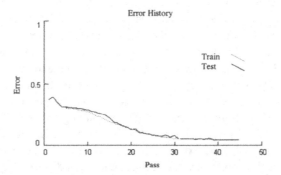

Figure 2.9. The graph shows training and testing errors for the multilayer error backpropagation neural network, as the network learns to predict the next value of a chaotic time series, given the current value. In this figure, the network has trained to criterion (less than 10% error for all testing exemplars) in about 45 passes through the training exemplars.

When the *Training: percent correct* value has reached 100%, network training will stop. At this point, the results matrix can be examined. There are four columns in this matrix: Columns 1 and 2 hold the target and network scores for the training exemplars; columns 3 and 4 hold the target and training scores for the testing exemplars. Columns 3 and 4 can be plotted to demonstrate how accurately the network was able to learn to predict the chaotic time series. This plot is shown in Figure 2.10.

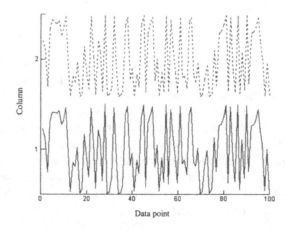

Figure 2.10. The bottom (solid) graph of column 3 of the network results matrix shows the target values that the network was asked to predict. These target values were generated using the chaotic logistic equation. Even though no apparent regularity exists in this time series, the network was nevertheless able to learn to predict these values accurately, as shown in the top (dotted) plot of column 4 of the network results matrix.

The network results matrix consists of 4 columns and 100 rows. Each row represents one exemplar. Columns 1 and 2 relate to the training exemplars. Column 1 shows the dummy codes that were assigned to the exemplar in the training file when the file was created. Column 2 of the network results matrix shows the corresponding value as predicted by the network. Columns 3 and 4 relate to the testing exemplars. Column 3 shows the target value from each testing exemplar. The network does not use this target value in any way. It is presented in the network results matrix solely for the convenience of the network user so that the performance of the network can be assessed. Column 4 shows the corresponding value that the network predicted for each of the testing exemplars. The degree of match between the values in columns 3 and 4 indicates the extent to which the network has learned the assigned task. We can see that, given any one data point generated by the logistic map, the network was able to accurately predict the next data point that the logistic map would generate.

Simulnet Exercise: Predicting Chaotic Data 2

In this exercise a backpropagation network will be asked to learn another chaotic system, the Lorenz flow. This system is related to the phenomenon of Rayleigh-Benard convection within a body of fluid receiving energy at the bottom and dissipating it at the top (for example, Baker and Gollub, 1990, p. 133). A simplified model of such convection dynamics was studied by Edward Lorenz (Lorenz, 1963) who developed a system of three coupled ordinary, nonlinear, differential equations:

$$x' = \sigma (y - x)$$

$$y' = Rx - y - xz$$

$$z' = xy - Bz$$

This highly simplified approximation has nevertheless been found to be capable of displaying a wide range of behaviors, from fixed points to limit cycles, to deterministic chaos, depending on the values assigned to the system parameters σ, R, and B. In these equations the prime denotes a differential with respect to time. These equations, therefore, describe how each of the system variables, x, y, and z, vary with time. In this system, x corresponds to the intensity of the convective movement and y corresponds to the temperature difference between top and bottom of the mass of fluid. The Lorenz system has been found to exhibit chaotic behavior for values of the three control parameters of $\sigma = 10$, $R = 28$, and $B = 8/3$. Figure 2.11 shows the resulting behavior of the Lorenz system, plotted on an xyz graph.

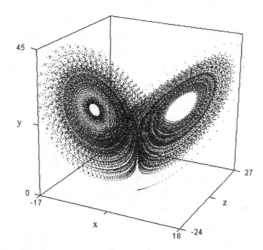

Figure 2.11. This figure shows the values generated by integrating the Lorenz flow for 4,000 steps, with a step size of 0.02, using a fourth-order Runge-Kutta integrator. The graph shows the characteristic double-spiral shape, technically a strange attractor. The behavior of the Lorenz system can be thought of as a point that moves within this three-dimensional space defined by the x, y, and z variables of the system. For the chosen parameter values, the point chaotically switches from one spiral to the other.

The following procedure was used to generate the network training and testing data. For each data set, the Lorenz equations were integrated for 1,000 time steps, using random starting values for x, y, and z, thus generating 1,000 data points for each of the system variables. Integration was carried out, using the *Model: Predefined Systems* function in Simulnet, which implements a fourth-order Runge-Kutta integrator, with a step size of 0.02. These 1,000 sets of data points were next converted into the network training matrix by stacking each pair of contiguous rows into 1 new row (using the *Stack rows* operation in the *Transformations* function, with the *number of rows* parameter set to 2). The resulting matrix consists of 500 rows of 6 columns, with each of these rows representing a single exemplar. In each row, columns 1 through 3 represent the values of x, y, and z at 1 time point, and comprise the predictor portion of the exemplar. Columns 4 through 6, in the same row, represent the values of these variables in the immediately following time point and comprise the criterion portion of the exemplar.

$$
\begin{array}{c}
\begin{array}{cccccc}
1 & 2 & 3 & 4 & 5 & 6
\end{array} \\
\begin{array}{c}
\textit{exemplar 1} \\
\textit{exemplar 2} \\
\vdots \\
\vdots \\
\textit{exemplar 500}
\end{array}
\left[
\begin{array}{cccccc}
x(t=1) & y(t=1) & z(t=1) & x(t=2) & y(t=2) & z(t=2) \\
x(t=3) & y(t=3) & z(t=3) & x(t=4) & y(t=4) & z(t=4) \\
& \vdots & & & & \\
& \vdots & & & & \vdots \\
x(499) & y(499) & z(499) & x(500) & y(500) & z(500)
\end{array}
\right]
\end{array}
$$

With this training matrix the network will be asked to learn to predict the set of values that would be generated by the Lorenz equations, given the previous values generated by the equations. The network testing matrix was constructed in exactly the same way. The initial values for x, y, and z were however chosen to be new, random values. As the network is trained, validation or testing will be periodically carried out by applying the testing data once every five training iterations.

Procedure

1. Close all forms on the desktop. From the *Network* menu, select the *BackProp Network* option. On the *BackProp Network* dialog form, click the *Load setup* button. On the *Load Setup* dialog form select file *lorenznn.set*. This file contains the setup values needed to run this exercise. In particular, a training file, *lorenz.trn*, and a testing file, *lorenz.tst*, will be loaded. Preset values for network parameters will be installed as well.

2. On the *BackProp Network* dialog form, click the *Train* button. A number of objects will appear on the desktop. These include a status form showing the current network training and testing errors and the pass number, a matrix containing network results, a graph that will show a record of training and testing errors averaged over the corresponding exemplars, a color-coded display of network weight values, and a histogram of the distribution of weight values. The network will then start training.

    As the network trains, training and testing error should eventually start to decrease, and training and testing correct percentages should rise. These figures are the percentages of the training and testing exemplars for which the network has achieved a score less than or equal to the criterion level. This criterion level is specified as 0.1 on the *Options* dialog form. In other words, an exemplar is judged to have been scored correctly by the network when the value that the network generates for that exemplar is within 10% of the target value for that exemplar as specified in the training and testing files. The number of passes needed to train the network to criterion cannot be determined precisely, but it will probably take between 50 and 200 passes.

    When the *Training: percent correct* value has reached 100%, network training will stop. At this point, the results matrix can be examined. There are 500 rows of 12 columns in this matrix. Columns 1 through 3 contain target scores for the training exemplars while columns 4 through 6 contain the corresponding network predicted values for the training exemplars. Columns 7 through 9 contain target scores for the testing exemplars, and columns 10 through 12 contain the corresponding network predicted values. Each row corresponds to a single exemplar. The degree to which the network has been able to learn the Lorenz system is indicated by the similarity between the target and predicted values for each of the three system variables. For evidence of this learning, we can examine the contents of columns 7 and 10, the target and predicted x values; columns 8 and 11, the target and predicted y values; and columns 9 and 12, the target and predicted z values. These columns are plotted in Figure 2.12.

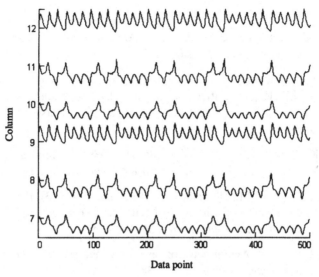

Figure 2.12. Columns 7 through 9 show the target values that the network was asked to predict: the x, y, and z variables, respectively, of the Lorenz equations. Columns 9 through 12 show the corresponding network predicted values for each of these variables. The network was able to learn the Lorenz system, as demonstrated by the similarity within column pairs 7 and 10, 8 and 11, and 9 and 12.

References

Angus, John E. (1991). Criteria for choosing the best neural network: I. *US Naval Health Research Center Report*, Rpt No 91–16 25.

Baker, G. L. and Gollub, J. P. (1990). *Chaotic Dynamics, an introduction*. Cambridge: Cambridge University Press.

Baum, E. and Haussler, D. (1989). What size net give valid generalization? *Neural Computation*, 1(1): 151 - 160.

Bioch, J. C., Verbeke, W., and van Dijk, M. W. (1994). Neural networks: New tools for data analysis? In P. J. G. Lisboa and M. J. Taylor (Eds.), *Proceedings of the Workshop on Neural Network Applications and Tools*, Los Alamitos, CA: IEEE Computer Society Press: 29-38.

Flotzinger, D., Kalcher, J., and Pfurtscheller, G. (1993). Suitability of Learning Vector Quantization for on-line learning: A case study of EEG classification. *Proceedings of the World Conference on Neural Networks (WCCN - 93), Vol. 1*. New Jersey: Lawrence Erlbaum Associates.

Funahashi, K. (1989). On the approximate realization of continuous mappings by neural networks. Neural Networks, 2: 183-192.

Gabor, Andrew J. and Seyal, Masud (1992). Automated interictal EEG spike detection using artificial neural networks. *Electroencephalography & Clinical Neurophysiology, 83(5)*: 271-280.

Grozinger, M., Kloppel, B., and Roschke, J. (1993). Recognition of rapid-eye movement (REM) sleep by artificial neural networks. *Proceedings of the World Conference on Neural Networks (WCCN - 93), Vol. 1*. New Jersey: Lawrence Erlbaum Associates.

Hornik, K., Stinchcombe, M., and White, H. (1989). Multilayer feedforward networks are universal approximators. *Neural Networks, 2(5)*: 359-366.

Jando, G., Siegel, R. M., Horvath, Z., and Buzaki, G. (1993). Pattern recognition of the electroencephalogram by artificial neural networks. *Electroencephalography and Clinical Neurophysiology, 86*: 100–109.

Kloppel, B. (1994a). Neural networks as a new method for EEG analysis. *Pharmacoelectroencephalography, 29*: 33-38.

Kloppel, B. (1994b). Application of neural networks for EEG analysis. *Pharmacoelectroencephalography, 29*: 39-46.

Kohonen, T. (1989). *Self Organization and Associative Memory*, 3rd Edition. New York: Springer-Verlag.

Lisboa, P. J. G., Mehridehnavi, A. R., and Martin, P. A. (1994). The interpretation of supervised neural networks. In P. J. G. Lisboa and M. J. Taylor (Eds.), *Proceedings of the Workshop on Neural Network Applications and Tools*, Los Alamitos, CA: IEEE Computer Society Press: 11–17.

Lorenz, E. N. (1963). Deterministic non-periodic flow. *Journal of Atmospheric Science, 20*: 130–141.

Minsky, M. and Papert, S. (1969). *Perceptrons*. Cambridge, Mass.: MIT Press.

Pfurtscheller, G., Flotzinger, D., Mohl, W., and Peltoranta, M. (1992). Prediction of the side of hand movements from single-trial multi channel EEG data using neural networks. *Electroencephalography & Clinical Neurophysiology 82(4)*: 313-315.

Rosenblatt, Frank (1958). The perceptron: A probabilistic model for information storage and organization in the brain. *Psychological Review, 65*: 386-408.

McClelland, J. L., Rumelhart, D. E., and the PDP Research Group (1986). *Parallel Distributed Processing: Explorations in the Microstructure of Cognition. Vols. 1 and 2*. Cambridge, Mass.: MIT Press.

Sietsma, Jocelyn; Dow, Robert J. (1991). Creating artificial neural networks that generalize. *Neural Networks, 4(1)*: 67-79.

Silva, F. M. and Almeida, L. B. (1990). Acceleration techniques for the backpropagation algorithm. In L. B. Almeida and Wellekens (Eds.) *Neural Networks, Europe Lecture Notes in Computer Science*. Berlin: Springer-Verlag: 110–119.

Slater, J. D., Wu, F. Y., Honig, L. S., Ramsay, R. E., and Morgan, R. (1994). Neural network analysis of the P300 event-related potential in multiple sclerosis. *Electroencephalography and Clinical Neurophysiology, 90*: 114-122.

Thornton, C. J. (1992). *Techniques in Computational Learning*. London: Chapman and Hall.

Vogl, T. P., Manglis, J. K., Rigler, A. K., Zink, W. T., and Alkon, D. L. (1988). Accelerating the convergence of the back-propagation method. *Biological Cybernetics, 59*: 257-263.

Webb, A. R. and Lowe, D. (1990). The optimized internal representation of multilayer classifier networks performs non-linear discriminant analysis. *Neural Networks 3(4)*: 367-375.

Werbos, P. J. (1974). *Beyond regression: New tools for prediction and analysis in the behavioral sciences*. Ph.D. dissertation, Harvard University, Cambridge, Mass.

Genetic Algorithms and Neural Networks

Introduction

The idea of applying principles of genetics to the evolution of mathematical structures, not of biological organisms, was developed by John Holland (1975). This idea has been termed the genetic algorithm approach. In this approach a population of number strings, or vectors, is created initially by randomly sampling from some distribution. Any one of these vectors is thus a set of random numbers, each of which can range between some limits. In an analogy with a population of biological entities, each vector is, in a limited sense, treated as an organism within an environment. This environment is the search space of the problem that is to be solved.

Before continuing with this description of genetic algorithms, the term search space will be described. Recall the discussion of weight space concerning neural networks trained by error backpropagation. The dimensions of the weight space in that discussion were the network weights, and any point in weight space corresponded to the state of the neural network with a particular set of weight values. The task of the error backpropagation algorithm was to find the point along the error surface in this weight space that corresponded to the minimum value of training error.

The notion of weight space can be generalized to the concept of a search space. A search space is simply the space, or hyperspace if more than three dimensions are involved, whose dimensions are the parameters that describe some system, such as a problem to be solved. Any point in the search space then corresponds to one particular set of values of the parameters of the problem. Solving the problem amounts to finding the particular point in the search space that corresponds to the desired outcome. In the case of the error backpropagation algorithm, the desired outcome was the lowest point of the error surface. A search space similarly contains a surface, whose topography represents the value of the problem outcome, for all combinations of the problem parameters.

Returning to the discussion of genetic algorithms, each of the vectors that comprise the population is essentially treated as a potential solution to the problem at hand. This population of vectors is then evaluated for fitness within the environment of the search space. This evaluation is carried out using some evaluation function that is appropriate to the particular problem. The fitness of one of these vectors is typically defined as the ability of the vector to locate the most rewarding regions of the search space. That is, the fitness of a vector is proportional to how good a solution to the problem that vector represents. A vector's fitness is thus a measure of how well the information encoded within the vector represents a solution to the given problem.

The process of evolving a solution to a problem in this way involves a number of operations that are loosely modeled on their counterparts from genetics. The most widely used genetic operators are mating, crossover, and mutation.

Genetic Operators

Modeled after the processes of biological genetics, pairs of vectors in the population are allowed to "mate" with a probability that is proportional to their fitness. The mating procedure typically involves one or more genetic operators. The two most commonly applied genetic operators are crossover and mutation.

The crossover operator, in analogy to the gene crossover phenomenon in cell reproduction, allows information within the vector population to be redistributed. In the crossover operator, a pair of mating vectors exchanges information by exchanging a subset of their components. The result is a new pair of vectors, each of which carries components from both of the parent vectors. This new pair of vectors then replaces the parent vectors in the population.

Two parameters are involved in the crossover operator, probability and location. The value of the probability of crossover parameter can be specified by the user. The second parameter, the location of the crossover point, is the point at which the two vectors are cut to form the subsections that are then exchanged. To illustrate the process of crossover, the pair of vectors **a** and **b**

$$\mathbf{a} = [2, 5, 1, 7, 4, 3, 8]$$

$$\mathbf{b} = [7, 9, 3, 2, 6, 5, 1]$$

are transformed by applying the crossover operator at the third digit position, to produce the vectors **a'** and **b'**:

$$\mathbf{a'} = [2, 5, 3, 2, 6, 5, 1]$$

$$\mathbf{b'} = [7, 9, 1, 7, 4, 3, 8]$$

Note that after the second digit (in this example) the two vectors have exchanged the components shown in italic.

The second operator, mutation, causes random changes in the components of the vectors in the new population. A mutation probability parameter controls the probability with which a mutation occurs. The value of this parameter can be specified by the user. Typically, mutation probability is set to a value that is approximately equal to the inverse of the population size. As an example of the operation of the mutation operator, the vector **c**

$$\mathbf{c} = [5, 8, 1, 4, 2, 9, 3]$$

is mutated by altering the value of the fifth component of the vector to produce the vector **c'**:

$$\mathbf{c'} = [5, 8, 1, 4, 6, 9, 3]$$

When all vectors in the parent population have been mated and have created replacement child-vectors, a new population has been created. This new population is in turn evaluated for fitness, mated and operated on, creating the subsequent generation. Figure 2.13 illustrates schematically the genetic algorithm approach.

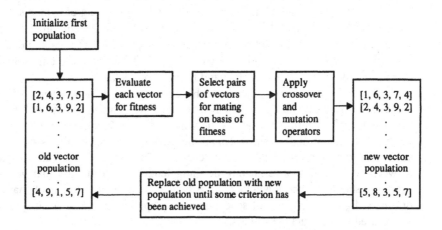

Figure 2.13. This schematic diagram of a genetic algorithm shows the functions that are carried out in each generation. Over a number of such generations the initial population is evolved to the point where it can meet some criterion with respect to the problem at hand.

What Makes Genetic Algorithms So Useful?

Genetic algorithms are capable of finding solutions to problems that are difficult for other older search techniques, such as the simplex algorithm, and even neural networks.

Difficult problems in this context include those whose search spaces contain many local minima. Recall that local minima are regions in the search space that correspond to solutions that are almost equally as good, but that are not the best solution. Backpropagation-trained neural networks, for example, are a class of search techniques that are at least potentially prone to entrapment by such local minima. Other, more traditional search procedures such as the simplex algorithm, are more than just potentially prone to the local minimum problem.

The power of genetic algorithms in finding solutions to such difficult problems is the result of the key property of genetic algorithms as demonstrated by John Holland (1975): Successive generations of vectors can contain exponentially increasing numbers of vectors in the best regions of the search space. This will be the case particularly when vector evaluation is carried out in such a way that short subsections of a vector, termed schemata, are sufficient in themselves to lead to a high value of fitness for that vector.

Before continuing with the main stream of this discussion of genetic algorithms, we will digress briefly to explain the meaning of the term schema (plural, schemata). Schemata, also referred to as similarity templates (Goldberg, 1989), are templates that describe the pattern of values within a vector or portion of a vector.

To illustrate this point, imagine that the vectors are constructed using only the digits 0 and 1 of binary notation. One such vector, for instance, might be the string of numbers [1, 0, 0, 1, 0, 1, 1, 1]. A schema could be defined as the string of symbols *1*1, where the symbol * is introduced to stand for 'don't care'. This schema thus represents any string of 0's and 1's that matches this particular template. Thus for example, the string 1, 1, 0, 1 matches the schema, while the string 1, 0, 1, 0 does not. The term schema is used in referring to, or comparing, vectors and portions of vectors.

Next, we introduce the concept of the defining length of a schema. The defining length of a schema is simply the distance between the first and last symbol in the schema that is not a *. The defining length of the schema *1*1 is 2; the distance between the first 1 in the schema, at digit position 2, and the second 1, at position 4, is 4 - 2 = 2. In the same way, the defining length of the schema 1*01*11* is 6. We can now distinguish between schema of short and long defining lengths, a distinction that will be of use shortly. It should be mentioned that while the vectors used in this example have consisted of binary digits, the concept of schemata can be applied no matter what system of notation is used to construct the vectors.

We return now to the discussion of what it is that makes genetic algorithms so useful. Ideally, we would like to be able to encode potential solutions within the vectors in such a way that good partial solutions are contained within short regions of the vector. That is, the vectors should be constructed in such a way that good solutions to the problem at hand can be constructed by assembling schema of relatively short defining length. Such schema will be referred to as building blocks.

On the other hand, we want to avoid the situation where a good solution is encoded over most of the length of the vector. Intuitively, schema of short defining length are less likely to be disrupted by the crossover operator, and thus more likely to survive through the mating process. With vectors whose fitness depends on short schema, the crossover operator has a reasonable probability of exchanging good partial solutions between vectors, with a low probability of disrupting these partial solutions. In contrast, vectors whose high fitness level depends on longer schema, or on noncontiguous portions of the vector, are less likely to pass their information on to succeeding generations because of the higher probability with which these longer or noncontiguous portions are likely to be disrupted during the process of mating.

In sum, the information within a short schema that represents a good partial solution to the problem at hand can be distributed throughout the population with a relatively low risk of disruption by crossover or mutation.

The significance of this effect was demonstrated by Holland (1975) in a result known as the *schema theorem*. Assuming that, among other conditions, good solutions can be assembled out of schema of short defining length, then in each succeeding generation the proportion of short and fit schemata will increase exponentially within the population. At the same time, the proportion of short and unfit schemata will decrease exponentially within the population. The overall result in terms of the problem search space will be, as stated earlier, that good regions of the search space will accumulate exponentially increasing numbers of vectors.

At the same time, the crossover and mutation operators work to ensure that all regions of the search space continue to be explored. The following statement describes the mechanism that enables genetic algorithms to deal with difficult problems, their relative immunity to being trapped in locally good regions of the search space: While the best regions, as defined by the fitness evaluation function, of the search space gather increasing numbers of vectors, all regions of the space continue to receive attention.

Approaches based on genetic algorithms in which a population of transformations is evolved over a number of generations, have been shown to be capable of searching complex search spaces with an efficiency that can surpass that of more traditional search methods used with distributed network models: such as, for instance, gradient descent methods like error backpropagation (Goldberg, 1989). The advantage that genetic algorithms possess in terms of their relative immunity to the local minimum problem may be offset, however, by disadvantages such as longer computation times on traditional serial computers. Fortunately, the ever-increasing availability of powerful desktop computers and workstations does serve to mitigate this drawback.

The Evaluation Function

A key consideration in the application of genetic algorithms to a particular problem is the design of the evaluation function. As an example of the design of an evaluation function, let us consider a typical application for a genetic algorithm—the classification of data vectors into a number of categories. In concrete terms such data vectors might, for instance, represent time series data recorded within one or another condition of an experiment.

Let us consider the situation where a genetic algorithm is used to classify such vectors into two categories. One possible evaluation function for this task would be to form the vector dot-product between a population vector and each of the data vectors in turn, from one of the categories. The average of this dot-product would then be computed and the procedure repeated for all of the data vectors in the two categories. The fitness of the population vector is then computed as the difference of the average dot-product for the first category and the average dot-product for the second category. A population vector with a high fitness would form a large average dot-product with the first category data vectors and, simultaneously, a small average dot-product with second category data vectors.

Using the Genetic Algorithm

A genetic algorithm is applied in much the same way as an error backpropagation-trained neural network: The genetic algorithm is put through a training stage in which a population of potential solution vectors is evolved, followed by a testing stage in which the fittest vector is selected for testing.

In the training stage, the genetic algorithm is evolved to convergence. Convergence is defined as the point at which the average fitness level of the population appears to level out. A level of fitness is computed, in turn, for each vector in the population. Vectors then mate with a probability proportional to their level of fitness. Genetic operators, such as cross-over and mutation, are applied creating a new population of vectors, which then replaces the previous population. This process continues until some criterion is reached, at which time the fittest vector in the final population is identified.

In the testing stage this fittest vector could be used, for example, as a classification rule able to discriminate between novel data vectors into the two categories. This classification procedure would be similar to that used in the training stage; a dot-product is formed between the classification rule and each test vector. This dot-product is then defined as the classification score for that test vector.

For a review of the application of genetic algorithms to a range of biomedical problems, see Levin (1995).

Creating a Neural Network-Genetic Algorithm Hybrid

As discussed earlier, genetic algorithms are able to deal with a potential limitation of the neural network approach to data analysis and prediction. Neural networks can potentially become trapped by a solution that may be better than a sample of alternative solutions, but which, at the same time, is not the best possible solution. Typically, a neural network is asked to find a pattern within a body of data. The data might, for instance, contain the values of a number of economic indicators (the independent variables), along with the value of a currency (the dependent variable). The problem in this example would then be to find the relationship between the independent variables, the status of the economic indicators, and the dependent variable, the value of the currency. The network is expected to find such a relationship, if one exists. The network may, however, converge upon a pattern within the data that might not represent the most significant relationship between the economic indicators and the currency value. Such a pattern is a local minimum of the search space for this particular problem, while the ideal goal is to find the global minimum, the overall best solution. Still another potential limitation of a neural network is that if the search space contains significant flat or nearly flat regions, the neural network may either fail to converge at all, or converge only very slowly.

One way of getting around these potential limitations, while retaining the benefits of the neural network as a universal function approximator, is to create a neural network-genetic algorithm hybrid. Essentially, the genetic algorithm allows the neural network to keep exploring all possible solutions until the best solution, given the available data, has been found. One way in which the genetic algorithm approach can be applied to a neural network is to create a population of neural networks, each of which is then allowed to independently train on the same data. To relate this idea to the previous discussion of population vectors, the structure of a network is encoded as a string of numbers. In a straightforward application, each string of number, or vector, represents all of the network weights. Knowing all of the weight values of a network completely defines a network, keeping constant the actual structure of the network; that is, the number of input, hidden, and output nodes. As an example, a genetic algorithm may be designed to evolve a network with two input nodes, one output node, and a single hidden level containing two hidden nodes. This network would then contain six weights. This network could be encoded as a population vector with six components, as shown in Figure 2.14.

Network structure:

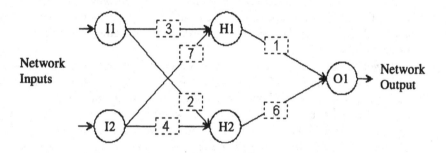

Network structure represented as a population vector: [3, 7, 2, 4, 1, 6]

Figure 2.14. The network containing six weights is encoded as a population vector with six components, each representing one of the network weights. In this illustration, weights are represented in decimal notation. With binary notation, each weight would be encoded as a number of components.

More elaborate applications can involve encoding not only network weights, but also the network configuration itself. Along with weight values, the number of hidden units, and even the number of hidden layers, could be encoded within a population vector.

Training of each of the networks in the population proceeds in exactly the way that was described previously. To give each network a chance to demonstrate how well it can learn, a network is allowed to train for some number of learning trials, or passes, through the network exemplars. That is, the entire set of training exemplars is presented to a network several times over, with network weights updated after each pass through the exemplars. Rather than choosing a fixed number of trials for all networks, the actual number of trials is allowed to vary randomly around some mean value chosen by experience or experiment. A unique value for the number of trials is chosen in this way for each network in the population.

Following the training period each network is tested. On the basis of the test results the network is assigned a fitness score. The fitness score is used to generate the next generation of networks. The new generation is created by applying the genetic operators, crossover and mutation, to the parent population. Each member of the parent population contributes information to the new population in proportion to its fitness score.

Thus, highly fit networks, those found to have achieved high test scores, contribute proportionally more information, in terms of network weights and network structure, to the new population than do less-fit networks. The function of the crossover operator is to redistribute information present in the newly created networks. The mutation operator is intended to provide a way of introducing a controlled degree of randomness or noise into the process of creating the new population of networks. Crossover allows pairs of networks in the new population to exchange portions of their respective structures. As Holland (1975) has shown, such redistribution of information results in an ever-increasing proportion of highly fit networks in successive generations. Mutation alters, by a random amount, a randomly selected portion of the structure of some of the newly created networks. The intent of this operator is to encourage the population of networks to keep exploring all regions of the search space, while accumulating networks in the most rewarding regions of the space.

Together, these features combine to alleviate the tendency encountered with the pure neural-network approach of settling into local rather than global minima of the search space. In practice, as mentioned previously, such local minima do not appear to be prevalent in the search spaces of most real-world problems: In a multi'dimensional hyperspace, a search procedure such as a neural network would have the opportunity to escape the local minimum by traveling along any of the multiple dimensions of the space. The higher the dimensionality of the search space, the more opportunities for escaping the local minimum. A true local minimum in a highly dimensioned search space would be one that was a minimum in terms of all the dimensions of the space. Again, as dimensionality increases, such local minima should be encountered less frequently.

The choice of which approach to use on a particular problem then is probably best found by experiment. A single neural network should generally converge to a low training error relatively quickly, while the neural network-genetic algorithm hybrid is less prone to entrapment by local minima. Without prior knowledge about the topography of the search space, and given sufficient time and computing resources, it is probably best to try both approaches on, at least, a subset of the available data, in order to gain some sense of the topography of the search space for the particular problem.

The Genetic Network Function in Simulnet

In this section, the various controls and settings that are available with the Simulnet *Genetic Network* function are listed and described. The *Genetic Network* function allows a population of multilayer feedforward neural networks to be evolved within the search space of the task at hand. Table 2.2 lists and describes a subset of the available controls and settings located on the *Genetic Network* and the *Genetic Network Options* dialog forms. This table does not include those controls that are also implemented on the *Back-Prop Network* dialog form, discussed earlier.

Before attempting to use the *Genetic Network* function, the reader is strongly encouraged to run Simulnet, select the *Genetic Network* function from the *Network* menu to show the *Genetic Network* dialog form, and then study the information in the following table. Going through this exercise is particularly important if the reader wishes to use this function with data other than that which has been supplied with Simulnet.

Table 2.2 *Genetic Network* **Controls**

Controls Located on *Genetic Network* dialog form

Control	Description
Network Architecture	
Population size	The population size parameter sets the size of the population of neural networks. As a rough rule-of-thumb, population sizes smaller than about 50 members do not really engage the power of the genetic algorithm approach. Small population sizes essentially amount to a brute-force search through the population for the best network generated by chance during initialization. On the other hand, large populations can require significant amounts of memory and computation time.
	One approach that may be useful is to start with a population of one, a single network. The optimum parameters for this network can be found relatively quickly. Alternatively, the *BackProp Network* function can be used for this purpose. In this way parameters, such as the minimum required number of hidden units, can be ascertained. This single, trial, network will presumably be seen to converge to a relatively low, but not zero, value of training error. At this point, the *Genetic Network* function can be used to create and evolve a population of such networks, out of which should emerge a single network that can outperform the original trial network.
Network Archiving	
Saving a network	While the saving a network function is also implemented in the *BackProp Network* function, it deserves mention here because of the way that the network to be saved is chosen.
	In each generation, during the evolution of the population of networks, the best network of the generation is identified. It is

this network that is used when testing is carried out during ongoing training. And it is this network that is selected for further training when, as training error falls below some specified level, training switches from genetic to error backpropagation mode. When the *Save Net* function is invoked, it is this network that is saved to disk.

Recreating a
network

A network that has been saved as a disk file can later be recreated using the *Load Files: Net* function. Given this capability, network training can be stopped at any point and the network saved to disk. The network can then be recreated at any future time and training resumed, either using the *Genetic Network* function, or using the *BackProp Network* function.

Note that if a network is recreated in this way for further training in the *Genetic Network* function, training should be carried on in the error backpropagation mode, rather than the genetic mode. To ensure that this happens, the value of the *Train: Switch* parameter should be set to one. Training will then be carried out in error backpropagation mode only if training error is greater than one (while it will not be, given a trained network).

Saving a
population

A population configuration can be saved to disk at any time using the *Save Pop* function.

A population configuration consists of the weight values for all networks in the population, along with a header specifying the population configuration: the size of the population; the number of input, hidden, and output nodes; and number of hidden layers in each network.

Recreating a
population

A population that has been saved as a disk file can later be recreated using the *Load Files: Pop* function. Given this capability, population evolution can be stopped at any point and the population saved to disk. The population can then be recreated at any future time and training resumed.

Display Objects

Status

The *Display: Status* option shows the number of the current training iteration and the *RMS* and *AVErage* training and testing errors, updated every training and testing iteration. The *Correct* indication on this display represents the percentage of output nodes with achieved values that are within the respective criterion level of the respective target outcome values in the training and testing sets.

There are a number of differences between the *Status* display in this function, and the *Status* display in the *BackProp Network* function. In the *Genetic Network* function, the values of training error shown on the *Status* form are average values computed over the entire population. Testing errors represent the errors for the best network in the population in each generation.

Error History

The *Display: Error History* option shows a graph of AVErage training errors (color is default plot color 1) and AVErage testing errors (color is default plot color 2) from the start of training. The number of iterations per graph is specified by the value of the *Passes* parameter. When this number of iterations has been graphed, the graph is cleared and restarted beginning with the next iteration.

As indicated in the note about the *Status* display, training error is computed over the population while testing error is computed for the best network in each generation.

Data Matrix

The *Display: Data Matrix* option shows a matrix containing target and predicted values for each of the criterion variables, in each exemplar, in both training and testing matrices. These values are arranged in four groups of columns, with one row for each exemplar. If n is the number of criterion variables, columns 1 to n contain the target values from the training set. Columns n + 1 to 2n contain the corresponding predicted values. Columns 2n + 1 to 3n contain the target values from the testing set. Columns 3n + 1 to 4n contain the corresponding predicted values. For instance, if there are two criterion variables, the first two columns contain the target training values, and the next two contain training predicted training values. Columns 4 to 6 contain target testing values, and columns 6 to 8 contain predicted testing values.

There are differences in the contents of this matrix between the *BackProp Network* and *Genetic Network* functions. In the *Genetic Network* function, the values in the second group of columns, those representing predicted training values, are averages computed over the population. The values in the fourth group of columns, showing predicted testing values, are computed using the best network of the current generation.

Weights

The *Display: Weights* option displays a color-coded representation of the weight values in the current population. Weight values are arranged as a matrix with the weights in one network represented by columns, and networks themselves represented by rows. This display is updated every training iteration. The last column in the display represents the fitness level of the network in that row.

Training Mode

Mode switch level	The mode switch level parameter determines the level of average training error (the average over the population of the average training error) below which training switches from genetic mode, in which the population of network is evolved, to error backpropagation mode, in which the current best network is selected for individual training using error backpropagation.
Number of trials per network	The number of trials per network parameter sets the number of iterations of error backpropagation training that each network in the population is given when training is being carried out in the genetic mode. For instance, if this parameter is set to 10 then during genetic training each network in the population is given 10 iteration of error backpropagation training through which the network can demonstrate how well it can learn.

Reinitialization

Reinitialize population	Reinitialize population, invoked using the *Restart* button, reinitializes the population using pseudo-random noise with characteristics set using the selected options for *Weight: Distribution* and *Weight: Range*.

Controls Located on *Genetic Network Options* dialog form

Training Parameters

Elitist Mode	The fittest network in every generation is retained and reinserted into the following generation. This option generally improves training convergence on problems of low-to-moderate difficulty. On difficult problems (for example, variations on the classic XOR problem), elitist training generally slows the rate of training convergence.
	If this option is not selected, all networks are treated equally. In each generation, all networks are evaluated and assigned a fitness value in proportion to which these networks are allowed to 'mate' and 'reproduce' to form the subsequent generation.

Genetic Parameters

Crossover	Crossover sets the rate of mutation within the population. This

value is equal to the probability that a network weight will be changed to a new randomly selected value. There are two choices for this parameter, selected using the following options:

Crossover: Adaptive: The level of crossover is allowed to vary between members of the population; the value of crossover depends on the network fitness distribution.

Crossover: Set: The level of crossover is the same for all members of the population, and is equal to the value specified for this parameter.

Training Parameters	
Mutation	Mutation sets the rate of mutation within the population. This value is equal to the probability that a network weight will be changed to a new randomly selected value. There are two choices for this parameter, selected using the following options:
	Mutation: Adaptive: The rate of mutation is allowed to vary between members of the population; the level of mutation depends on the network fitness distribution.
	Mutation: Set: The rate of mutation is the same for all members of the population, and is equal to the value specified for this parameter.

Simulnet Exercise: Learning a Functional Relationship

This exercise will illustrate the procedure to be followed in order to use the *Genetic Network* function in Simulnet. This function evolves a population of multilayer neural networks that are each trained using the error backpropagation rule. In this exercise a population of 40 networks will be evolved. Each network has 24 input nodes corresponding to 24 independent variables in the training data, 2 hidden nodes in a single hidden level, and a single output node for the single dependent variable. In each generation, all networks are given the opportunity to demonstrate how well they can learn: In the present exercise each network is allowed 10 passes through the training exemplars. Each of the networks is then evaluated for fitness by being tested on how well they perform on a set of testing exemplars. When, over the course of successive generations, a single network eventually evolves that has a sufficiently low testing error, this network is selected from the population for further training using the backpropagation rule. The rest of the population, having served its purpose in creating this one network, is now discarded.

The genetic network function thus operates in two training modes. The first, termed genetic mode, is one in which the entire population is evolved. The second, termed backpropagation mode, is one in which the best network of a population, after a low enough testing error has been achieved, is selected for further training by itself.

The networks will be asked to learn a functional relationship between 24 independent variables and 1 dependent variable. The independent variables are sampled from a noisy sine wave. The single dependent variable is the value of the starting phase angle of the sine waves. The genetic network will be trained on 40 exemplars of such a noisy sine wave, with 10 exemplars for each of 4 phase angles; 0, 30, 60, and 90 degrees. The genetic network will then be tested on 7 exemplars, 1 for each of the following phase angles: 0, 15. 30, 45, 60, 75, and 90 degrees. The genetic network will thus need to learn the relationship between the independent variables and the associated phase angle in order to be able to successfully generate a prediction for phase angles on which it has not been trained: 15, 45, and 75 degrees. The accuracy with which the network can generate these interpolated values will be an indication of how well the network is able to generalize what it has learned. The pseudo-random noise component in each exemplar is independent of that in all other exemplars.

As it is being evolved, the population of networks will be periodically tested (once every five passes through the training data).

Procedure

1. Close all forms on the desktop. From the *Network* menu, select the *Genetic Network* option. On the *Genetic Network* dialog form, click the *Load setup* button. On the *Load Setup* dialog form, select file *phasegn.set*. This file contains the setup values that are needed to run this exercise. In particular a training file, *phase.trn*, and a testing file, *phase.tst*, will be loaded.

2. On the *Genetic Network* dialog form, click the *Train* button. A number of objects will appear on the desktop: A status form showing the current population training and testing errors and the pass number; a matrix form that will be used to hold the training and testing results; and a graph form on which will be plotted a record of the population training and testing errors, averaged over the corresponding exemplars. The population will then start evolving in genetic mode. When any one network in the population has achieved a sufficiently low value of testing error, training of the best network in that population will be trained further in backpropagation mode.

    As the population evolves, training and testing error should eventually start to decrease, and training and testing correct percentages should rise. These figures are the percentages of the training and testing exemplars for which a score less than or equal to the criterion level has been achieved. Criterion levels are specified on the *Options* dialog form as 0.2 for training error and 0.1 for testing error.

    In other words, a training exemplar is judged to have been scored correctly by the population when the average value that the population generates for that exemplar is within 20% of its target value. Correspondingly, a testing exemplar is judged to have been scored correctly by the best network in the current population network when the value that a network generates for that exemplar is within 10% of its target value.

The number of passes needed to train the network to criterion cannot be determined precisely, but it will probably take between 20 and 50 passes through the training exemplars. The graph showing the training and testing errors should resemble Figure 2.15. This graph, entitled Error History, contains two plot lines; one for training error averaged over all training exemplars, and one for testing error averaged over all training exemplars. The graph shows the network training to criterion in about nine passes through the training exemplars. In practice, however, it may take up to 50 passes to train the network.

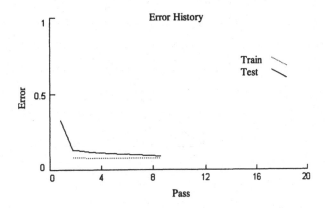

Figure 2.15. The graph shows training and testing errors for the genetic network as the population of networks learns the functional relationship in the training exemplars. In this figure, the network has trained to criterion (less than 10% error for all testing exemplars) in about nine passes through the training exemplars.

The networks trained in genetic mode for only one pass through the training exemplars. After this first pass, at least one network in the population achieved an average training error of less than the critical error of 0.4; the error level at which training mode switched from genetic to backpropagation mode. This switch value is set using the *Train: switch* data entry field on the *Genetic Network* dialog form. From the second pass onwards, training mode was backpropagation: The best network in the population was selected for training by itself.

When the *Training: percent correct* level for this best network has reached 100% training will stop. At this point the results matrix can be examined. There are 4 columns in this matrix: Columns 1 and 2 hold the target values and network estimates of the dependent variable for each of the training exemplars. Columns 3 and 4 hold the corresponding target values and training estimates for the testing exemplars. Since there are only 7 testing exemplars, only the first 7 rows of columns 3 and 4 have non-zero values. The other rows in columns 3 and 4 are not used. Columns 3 and 4 can be plotted to show how accurately the network was able to learn the functional relationship between the noisy time-series data and the value of phase angle. This plot is shown in Figure 2.16.

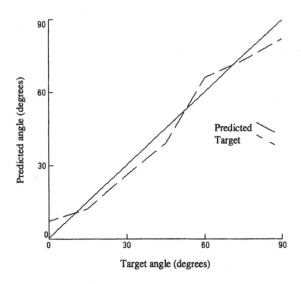

Figure 2.16. The solid line shows the target phase angle. The dashed line shows the predicted phase angle that the genetic network was able to achieve after training. Although not perfect, the relationship learned by the network is nevertheless an approximation to the true relationship, shown as a dotted line on the graph.

The genetic network results matrix consists of 4 columns and 40 rows. Each row represents one exemplar. Columns 1 and 2 relate to the training exemplars. Column 1 shows the value of phase angle that was assigned to each exemplar in the training file when the file was created. Column 2 of the network results matrix shows the corresponding value as predicted by the network. Columns 3 and 4 relate to the testing exemplars. Column 3 shows the target value of phase angle from each testing exemplar. The network does not use this target value in any way. It is presented in the network results matrix solely for the convenience of the network user so that the performance of the network can be assessed. Column 4 shows the corresponding value of phase angle that the network predicted for each of the testing exemplars. The degree of match between the values in columns 3 and 4, shown in Figure 2.16, indicates the extent to which the network has learned the assigned task, predicted the next value of the chaotic time series, given the current value.

References

Goldberg, D. E. (1989). *Genetic Algorithms in Search, Optimization, and Machine Learning.* Reading Mass.: Addison-Wesley.

Holland, J. H. (1975). *Adaptation in Natural and Artificial Systems.* Ann Arbor: The University of Michigan Press.

Levin, M. (1995). *Use of genetic algorithms to solve biomedical problems.* M.D. Computing, 12(3), 193-9.

The Probabilistic Network

Introduction

The operation of the *Probabilistic Network* function in Simulnet is based on the principles of the generalized regression neural network (GRNN) developed by Don Specht (Specht, 1991), and discussed by Wasserman (1993). The GRNN, like the error backpropagation neural network, is able to approximate any functional relationship between a set of independent and dependent variables. The following description will be based on a GRNN used as a classifier; that is, to learn to place test exemplars into one of two or more categories. The GRNN can, however, also function as an associator, learning the nature of the relationship between a set of predictor and criterion variables.

Structurally the GRNN somewhat resembles the multilayer backpropagation neural network that was discussed earlier (see Figure 2.17). The GRNN has a number of input nodes equal to the number of predictor variables. These input nodes of the GRNN, like those of a backpropagation network, are merely connection points to which the elements of the test exemplars are applied, one at a time. The GRNN also has a number of hidden units. Unlike the backpropagation neural network however, the number of hidden nodes is defined. The number of hidden nodes is, in fact, equal to the number of training exemplars; one hidden node is assigned to each training exemplar. Unlike the backpropagation network then, the GRNN does not require an estimate of the number of hidden units to be made before training can begin. Finally, if the GRNN is used as a classifier, the number of output nodes is equal to the number of categories being discriminated. More generally, the number of output nodes is equal to the number of dependent variables whose values are being predicted.

Input (test) vector

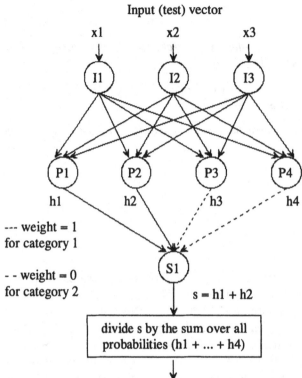

Input nodes I receive test vector

Input weights are components of each test vector

Pattern nodes P compute the probability of the test vector for each training vector

Output weights are 1 for category 1 and 0 for category 2

Summation node S1 sumes the probabilities for category 1

Figure 2.17. This example of a *Probabilistic Network* is designed to classify input vectors (test vectors) into one of two categories. To do this the network uses information about four training vectors, each stored in the weights connecting the input and pattern nodes. The output of the network is the probability that the currently applied test vector belongs to a particular category (labeled as category 1).

The test vectors, or test exemplars, that are applied to this network each consist of three components. The network is designed to classify these test exemplars into 1 of 2 categories, labeled 1 and 2. Each component of an input vector is applied to 1 of the input nodes i1 through i3. In the next level, the network contains 4 pattern nodes, p1 through p4, 1 for each of 4 training exemplars. The weights between the input nodes and these pattern nodes contain the actual values of the components of the training exemplars.

For example, the weights between all of the input nodes and pattern node p1 consist of the components of training vector 1, and so on for pattern nodes p2 through p4. As a result, each of the pattern nodes receives the vector dot-product of the currently applied test vector and the training vectors corresponding to that pattern node. Each pattern node in turn then computes the probability, h1 through h4, of the applied test exemplar given its particular training exemplar. Thus, pattern node p1 computes the probability of the current test exemplar given the probability density function of training exemplar 1. The 4 training exemplars belong to 1 of the 2 categories, 1 and 2; training exemplars 1 and 2 belong to category 1, while training exemplars 3 and 4 belong to category 2. Weights between the pattern nodes and the summation node s1 are designed to allow that summation node to compute the sum over the probabilities that correspond to only 1 of the categories, in this case category 1.

This summation node thus sums the probabilities h1 and h2. The last step is to convert this sum of probabilities into a true probability p. The sum of probabilities is divided by the sum over all probabilities (h1 + h2 + h3 + h4):

$$p = (h1 + h2) / (h1 + h2 + h3 + h4)$$

Probability p now represents the probability that the currently applied test exemplar belongs to category 1. The probability of this test exemplar belonging to category 2 is then just 1 - p. For this reason, if exemplars are being classified into two categories, only a single output node is required. Typically, if exemplars are being classified into a number of categories, the number of output nodes that is required is one less than the number of categories. Economizing on output nodes in this way will decrease computation time.

Algorithm

The following algorithm describes the testing stage of the GRNN.

For each test exemplar x_i

1. For each training exemplar u_j

    1.1 Compute the probability of x_i given the probability density function of u_j:

    $$h_j = Exp[- (x_i - u_j)^T(x_i - u_j) / (2\sigma^2)]$$

2. Compute the sum over the probabilities for all training exemplars: $\sum h_j$

3. For each output (category) k

    3.1 Compute the sum of the probabilities h_j for training exemplars from category k:

    $$s_k = \sum h_{j-k}$$

    3.2 Convert this sum to a probability by dividing s_k by the sum over all h_j:

    $$p_k = [\sum h_{j-k}] / [\sum h_j]$$

where T denotes the transpose operation; σ is smoothing, the width of the probability density function; and p_k is the probability of test exemplar x_i given all training exemplars h_j from category k.

Comparing the GRNN with the Error Backpropagation Network

While the GRNN may resemble an error backpropagation network structurally, the two types of networks differ functionally in a number of fundamental ways. First, with the GRNN there is no counterpart to the iterated network training stage. Instead, the entire training matrix is installed in the GRNN as the weights between the input and hidden layers. In more detail, the weights between the input nodes and each hidden node represent a single training exemplar. Thus, the weights between the input level and hidden node 1 are the components of the predictor part of training exemplar 1.

Recall that each exemplar, whether in the training matrix or the testing matrix, consists of two sections: The first section consists of the values of the predictor variables, the variables being used to predict some outcome; while the second section consists of the values of the criterion variables, the variables being predicted. The equivalent of training the GRNN thus takes no more time than is required to load the contents of the training file into working (RAM) memory. This scheme is in sharp contrast with that used with error backpropagation networks which interactively apply a heuristic, such as the method of steepest descent, to adjust the values of the input node to hidden node weights

The testing stage of the GRNN also differs significantly from that of error backpropagation networks. In order to describe the GRNN testing stage it is useful to first state what the outputs of the GRNN represent (again, for the purposes of this discussion, when the GRNN is used as a classifier). The outputs of the GRNN are the probabilities that the test exemplars belong to the categories being discriminated. The GRNN implements a procedure for estimating the probability of a test exemplar vector given a set of training exemplars, based on the principle of Bayesian classification. The GRNN will in fact approach an optimum Bayesian classifier given a large enough number of training exemplars (Wasserman, 1993). The algorithm used for GRNN testing can now be described as follows.

The testing stage begins with a testing exemplar being applied to the input nodes. Each hidden node will thus receive the product, and more precisely the vector dot-product, of the testing exemplar and the training exemplar corresponding to that hidden node. This vector dot-product is a direct measure of the collinearity, or in a general sense the similarity, between the test vector and a training vector. Other measures of similarity can also be used. The particular algorithm implemented within Simulnet uses the sum of squares of the difference between the test and training vectors.

Each hidden node then performs a nonlinear transformation on this dot-product. While in the error backpropagation network the transformation generally involves the logistic function. The corresponding transformation in the case of the GRNN involves the exponential function. The meaning of this nonlinearly transformed dot-product is that it represents the probability of obtaining the testing exemplar, given a probability density function with a mean equal to the training exemplar and standard deviation defined by a parameter referred to as smoothing. (Generally, smoothing is the only parameter that needs to be selected when using the GRNN.)

Straightforwardly, the GRNN computes at each hidden node the probability of the current test exemplar, given the existence of the training exemplar corresponding to that hidden node. In sum, the more similar the testing and training exemplars are (or in alternative terms, the more nearly collinear they are), the greater will be the resulting probability of that testing exemplar given the training exemplar.

These individual probabilities now need to be combined in order to generate the desired output of the GRNN; that is, the probability of the test exemplar given all of the training exemplars. This combining is performed in the hidden-to-output section of the GRNN. The outputs of each of the hidden nodes are connected to each of the output nodes. As in the error backpropagation network, these connections between the hidden and output nodes contain weights.

However, and again in contrast with the error backpropagation network, the weights in the case of the GRNN are not trained, but rather are assigned values. These values are dummy codes representing the category of each of the hidden nodes. Remember that each hidden node represents one training exemplar, and that exemplar belongs to one of the categories being discriminated. The dummy codes between a hidden node and all the output nodes are 1 for the output node that represents the same category as the training node, and 0 for all other output nodes. If there are two categories, A and B, being discriminated for example, the GRNN will have two output nodes, node A and node B. Assume that hidden node 1, representing training exemplar 1, belongs to category A. The weight between hidden node 1 and output node A will be 1, and the weight between hidden node 1 and output node B will be 0.

The effect of this coding is to connect only hidden and output nodes of the same category. The result is that an output node of a particular category will receive inputs only from hidden nodes of the same category. The output node then simply sums these inputs from the hidden nodes. While each of these inputs represents the probability of the current test exemplar given a particular training exemplar, this sum represents a measure related to the probability of the current testing exemplar given all of the training exemplars in one category. Finally, in order to generate an output that actually does represent the probability, the value at each output node is normalized by dividing it by the sum of all hidden node outputs.

Finally then, for this two-category exemplar, the value generated by the network at output node 1 is the probability that the currently applied test exemplar belongs to category 1. The value at output 2 is the probability that the testing exemplar belongs to category 2.

This technique of combining the probability density functions of individual exemplars of a category to approximate the probability density function of the category stems from work by Parzen (1962). Parzen showed that with a sufficient number of exemplars of a class, the result will approach the true probability density function of the category.

An advantage that the *Probabilistic Network* has over the neural network and the genetic network is the single-pass nature of the algorithm. Training and testing can typically be several orders of magnitude faster for the *Probabilistic Network* than for the neural or genetic networks. A potential limitation is that, since all training exemplars are stored in working memory (RAM), the size of the training data set is limited by the amount of available memory. With 4 Mb of extended memory, a training file can consist of up to roughly several thousand exemplars, with several hundred variables in each exemplar.

The Probabilistic Network Function in Simulnet

In this section, the various controls and settings that are available with the Simulnet *Probabilistic Network* function are listed and described. The *Probabilistic Network* function implements a universal function approximator, similar to the function performed by the backpropagation-trained neural network. A key distinction is that the *Probabilistic Network* involves a single-pass, rather than an iterated procedure.

Before attempting to use the *Probabilistic Network* function, the reader is strongly encouraged to run Simulnet, select the *Probabilistic Network* function from the *Network* menu to show the *Probabilistic Network* dialog form, and then study the information in the following table. Going through this exercise is particularly important if the reader wishes to use this function with data other than that which has been supplied with Simulnet.

Table 2.3 lists and describes the available controls and settings located on the *Probabilistic Network* dialog form. Descriptions of controls that are also found on the *Back-Prop Network* dialog form are described in the tables in the section for that function, and are not repeated here.

Table 2.3 Probabilistic Network Controls

Controls Located on *Probabilistic Network* dialog form

Control	Description
Training Data	
Training Files	The *Probabilistic Network* function allows for the selection of a single training file, or a set of training files. Such a set of training files might be generated if, for example, the user wants to find out how well a set of exemplars from several categories can be discriminated, using a moving window through the data. A separate training file can be created, using the *Create Training File* function, for each of several sets of rows in a set of exemplar files. These sets of rows can be thought of as a moving window through the data. Each such window, generating an individual training file, can be separately analyzed. By comparing the individual analyses, the degree of discriminability between categories being analyzed can be ascertained as a function of the window position in the data.

When a single training file is selected, network training is carried out using that file. Network testing can be performed using either the exemplars in the training file itself, the jackknife testing option, or using the exemplars in a separate testing file.

When multiple training files are selected, the *Jackknife test mode* option must be used, and is automatically selected. Thus, with multiple training files the testing must be carried out using the exemplars in the training files themselves.

Testing Parameters

Jackknifed operation	With a relatively small number of available network exemplars, random resampling can be used for network training and validation. Random resampling involves drawing at random and without replacement, a number of exemplars from the pool of exemplars. The network is trained on the exemplars remaining in the pool, and tested on the exemplars that were drawn. This procedure is repeated many times, a new random sample being drawn each time. A testing error rate is computed for each draw, and an overall network testing rate is computed by averaging over the individual rates.

When random resampling is carried out by withdrawing a single exemplar each time, the procedure is referred to as the *leave-one-out* method, or somewhat loosely, *jackknifing*. (While these two terms refer to methods that are virtually identical, the formal aims of the two are different.)

In this procedure each one of the exemplars is drawn from the pool, one at a time. The network is trained on all remaining exemplars, and tested on the single withdrawn exemplar. The testing error rate for that single exemplar is recorded and becomes the testing rate for that exemplar. This procedure is repeated for each of the exemplars in the pool in turn.

To access this function, select the *Testing Data: Jackknife test mode* option.

Smoothing — The *Probabilistic Network* develops a probability density function (PDF) for each of the training exemplars. The output of the network is the probability of getting a test exemplar, given the PDF's of all the training exemplars. Network performance is affected by the width of these PDF's. To explain the significance of width of the PDF's consider the *Probabilistic Network* as a classifier. In the limit of very wide PDF's, the decision boundaries between categories become hyperplanes. Correspondingly, categories separated by nonlinear boundaries cannot be effectively separated. As the width of the PDF's decreases, decision boundaries can become increasingly nonlinear. For very narrow PDF's, the network approaches a nearest neighbor classifier. Optimum performance generally requires some intermediate PDF width.

PDF width is determined by the value entered in the *Testing Parameters: Smoothing* field.

Simulnet Exercise: Data Classification

In this exercise the *Probabilistic Network* will be asked to learn to discriminate between exemplars of data that belong to either one of two categories. Used in this way, the *Probabilistic Network* functions as a classifier. Such a classification function has been found to be useful in a wide range of applications involving pattern recognition.

The exemplars used in this exercise were created by adding pseudo-random noise to sine and cosine functions. Each exemplar consists of 24 data points, together with 1 data point indicating the category to which the exemplar belongs. Category membership is dummy-coded as 0 for the noisy sine function exemplars, and 1 for the noisy cosine function exemplars. A total of 10 such exemplars, 5 sine and 5 cosine, has been assembled in the file *sincos.trn*. In order to test the performance of the network once it has been trained, a second file, *sincos.tst* has been assembled using 10 similar exemplars, but created using unrelated pseudo-random noise. As well, the noise component within each exemplar is independent of the noise component of any other exemplar.

Procedure

1. Close all forms on the desktop. From the *Network* menu, select the *Probabilistic Network* option. On the *Probabilistic Network* form, click the *Load Setup* button. On the *Load Setup* dialog form, select file *sincosnn.set*. This will load training file *sincos.trn* and testing file *sincos.tst*. A number of other setup parameters will be loaded as well, in particular a value of smoothing of 0.5.

2. On the *Probabilistic Network* dialog form, click the *Compute* button. The network may take up to several seconds to complete its operations. When the network has finished its computations, a matrix form will appear containing the results matrix. This matrix should contain the values shown in the following table (rounded to three significant figures).

Network Classification Scores	
Dummy code	**Network score**
0	0.0000785
0	0.126
0	0.000361
0	0.0202
0	0.00497
1	0.974
1	0.999
1	0.988
1	0.999
1	0.420

The network results matrix consists of 2 columns and 10 rows. Each row represents one exemplar in the testing file. Column 1 shows the dummy code that was assigned to each exemplar in the testing file when the file was first created. The first five exemplars are coded with a zero, indicating that they belong to the noisy sine function category. The second five exemplars are coded with a one, indicating that they belong to the noisy cosine function category. These dummy codes are shown in this network results matrix for the convenience of the network user only. The network does not use the codes in the testing file in any way.

Column 2 of the network results matrix shows the actual score assigned to each of the exemplars by the network. Again, this score is the probability that the corresponding exemplar belongs to the category. For example, the 6th exemplar (row 6 of the table) belongs to category 1 with a probability of 0.974.

Two features of these results are apparent. First, the network scores never exactly match the assigned dummy codes. Second, most of the scores are nevertheless close in value to the value of the dummy codes. This indicates that the network was indeed able to learn to effectively distinguish between the sine and cosine exemplars.

3. Set several values for smoothing in the range of 0.01 to 10, and run the network for each different value. Note how changing the value of smoothing affects the values in the results matrix.

Two general effects should be observed: First, network results change little unless relatively large changes are made in the *Smoothing* parameter. Generally, a change by a factor of two is needed to substantially change the network results. Second, as the value of smoothing is made smaller, the network scores become more extreme, tending towards the values of the dummy codes. On the other hand, as the value of smoothing is made larger, network scores become somewhat more evenly distributed between the extremes of zero and one.

Simulnet Exercise: Chaotic Time Series Prediction

In this exercise the *Probabilistic Network* will be asked to learn a functional relationship. The network will be used to predict the value of a chaotic time series, given the values of the preceding two data points. The chaotic time series is generated using the logistic map, defined by the difference equation

$$y_{n+1} = y_n + G(y_n(1 - y_n))$$

where G is the growth parameter and is set to a value of 3. In this application the *Probabilistic Network* will be acting as a function approximator in that it will be attempting to learn the principle of the logistic function. It happens that the value of the y variable generated by the logistic map depends only on the single immediately preceding value of y. Only one data value is actually required as the predictor. For the purpose of this exercise however, the preceding two values of the y variable will be used.

The following procedure was used to generate the training and testing data. For each data set, the logistic equation was iterated 300 times using a random starting value, thus generating a chaotic time series of 300 data points. These 300 data points were then formed into 100 sets of points. Each set consists of three contiguous points from the original time series. These 100 sets were organized as a matrix with 3 columns of 100 rows. In each row, columns 1 and 2 contain the first 2 data points of each set, while column 3 contains the third data point of each set. In terms of the original chaotic time series, within each matrix row the data points in column 3 immediately follow in time sequence the data points in columns 1 and 2, as follows:

$$
\begin{array}{l}
\textit{exemplar 1} \\
\textit{exemplar 2} \\
\quad \cdot \\
\quad \cdot \\
\textit{exemplar 100}
\end{array}
\left[
\begin{array}{ccc}
\text{time-point 1} & \text{time-point 2} & \text{time-point 3} \\
\text{time-point 4} & \text{time-point 5} & \text{time-point 6} \\
\cdot & \cdot & \cdot \\
\cdot & \cdot & \cdot \\
\text{time-point 298} & \text{time-point 299} & \text{time-point 300}
\end{array}
\right]
$$

Each row of this matrix constitutes one network exemplar: For each exemplar there are 2 independent variables, in columns 1 and 2; and 1 dependent variable, in column 3. The network will therefore be asked to predict the value that would be generated by the logistic equation given the preceding two values.

Two sets of 100 triples of data points were generated using this procedure. The training set is stored in file *log1.trn*, and the testing set is stored in file *log1.tst*. Each set was constructed to be independent of the other by starting the logistic equation each time with a new and random initial value.

Procedure

1. Close all forms on the desktop. From the *Network* menu, select the *Probabilistic Network* option. On the *Probabilistic Network* form, click the *Load Setup* button. On the *Load Setup* dialog form, select file *log1pn.set*. This will load training file

log1.trn and testing file *log1.tst*. A number of other setup parameters will be loaded as well, in particular a value of smoothing of 0.05.

2. On the *Probabilistic Network* dialog form, click the *Compute* button. The network may take up to several seconds to complete its operations. When the network has finished its computations, a matrix form will appear containing the results matrix. This matrix will consist of 2 columns of 100 rows. Each row represents one set of data values. Column 1 contains the target data values, while column 2 contains the actual data values as predicted by the *Probabilistic Network*. Columns 1 and 2 are graphed in Figure 2.18:

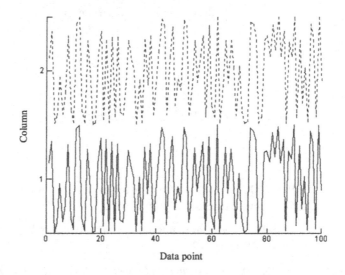

Figure 2.18. The solid line, the graph of column 1 of the network results matrix, shows the target data points that the *Probabilistic Network* was asked to predict. The dotted line, the graph of column 2 of the network results matrix, shows the actual data points as predicted by the network. The value of the smoothing parameter for these results was 0.05. A measure of the similarity of the two patterns of data points, Pearson product-moment correlation, was found to be 0.998, indicating a high degree of correlation: The *Probabilistic Network* was able to learn to accurately predict the chaotic time series.

3. The preceding results were obtained with the value of smoothing that was set as part of the setup in file *log1pn.set*. To examine the effect of different values of smoothing, repeat the network computation for each of the following values of smoothing; 2, 1, 0.5, 0.25, and 0.1. Target and predicted data values should become increasingly similar as the value of smoothing is decreased. To check for the degree of similarity between the target data in column 1 and the predicted data in column 2 of each matrix, graph the 2 columns with the *Attributes: Overlap* and the *Attributes: color* op-

tions selected on the *XY Graph* dialog form. With these options selected, the two columns will be graphed on top of each other in two different colors. Any differences between the two graphs should be easily apparent.

As a quantitative measure of the similarity between target and predicted data values, the correlation between columns 1 and 2 of the network results matrix will be computed using the *Correlations between columns* operation on the *Matrix Functions* dialog form. See the Simulnet Help facility for information about this function. The following table shows the resulting values of correlation for the different values of the smoothing parameter (rounded to three significant figures).

Prediction Performance	
Smoothing	**Correlation**
2	0.807
1	0.851
0.5	0.953
0.25	0.991
0.1	0.998
0.05	0.998

References

Parzen, E. (1962). On estimation of a probability density function and mode. *Annals of Mathematical Statistics*, 33, 1065-1076.

Specht, D. F. (1991). A general regression neural network. *IEEE Transaction on Neural Networks*, 2(6), 568-576.

Wasserman, P. D. (1993). *Advanced Methods in Neural Computing*. New York: Van Nostrand Reinhold.

The Vector Quantizer Network

Introduction

The *Vector Quantizer Network* function is founded on the principle of a vector quantizer, an algorithm that was designed for the task of data compression in signal transmission. A vector quantizer aims to represent a set of signal, or data, vectors using a relatively small number of reference, or codebook, vectors. In this way, a compressed representation of the data vectors is generated. Then, instead of transmitting the data vectors, their compressed representation, the codebook vectors, each representing a group of data vectors, can be transmitted instead. In order to accomplish this compression, a vector quantizer partitions, or 'quantizes' the space spanned by the signal vectors. Each partition represents one group or category within the set of signal vectors: Vectors lying within a partition are classified by the vector quantizer as belonging to the same group. Simulnet implements functions that are based on two variations on the theme of vector quantization; a LVQ (whose acronym is used for the name of this Simulnet function), and a SOM. The original algorithms of the LVQ and the SOM were developed by Teuvo Kohonen (1989).

The Self-Organizing Map

The SOM is designed to perform vector quantization, or classification, without the benefit of feedback. By definition then, the SOM belongs in the class of unsupervised learning algorithms. The SOM algorithm starts with a network of nodes, each of which is associated with one reference vector. Reference vector elements are initialized by drawing from a pool of pseudo-random values. The network is then trained by presenting it with a set of training exemplars. Exemplars are presented sequentially. Here, the term training vector refers to the predictor section, the set of independent variables, of each of the training exemplars.

As a result of this exposure to the training vectors, the network nodes self-organize to form groupings that reflect corresponding groupings in the network training data. In this process of self-organization, each of the reference vectors adopts the category labels of the training vectors to which they are most similar.

To describe this process in more detail, when a training vector is presented to the network, the most similar reference vector is identified. This reference vector, along with the vectors in some predefined neighborhood around it, is modified so that it still more closely resembles the training vector. These reference vectors then adopt the category label of the training vector. This process is repeated for each of the training vectors.

Reference vectors thus in a sense compete with each other to see which is closest to the currently applied data vector. From this behavior, the learning mechanism in this network can be described as competitive learning. As stated, when a training vector is presented, not only the most similar vector, but also vectors in some region surrounding the most similar vector, are modified to more closely resemble the applied training vector. Successive training vectors thus continually modify the array of network nodes: The winning vector in any one training iteration takes the opportunity to enlist a corps of surrounding vectors. While some of these surrounding vectors may already belong to the category of the applied training vector, others may have belonged to a different category. The currently applied training vector thus limits the sphere of influence of previously applied training vectors that represent other categories. This behavior can be viewed as a form of lateral inhibition: The winning vector inhibits, or at least curtails, the effects of previous vectors of dissimilar categories. The term lateral in this context simply means that the effect spreads from the locus of the winning vector to affect other vectors beside it.

Lateral inhibition is found widely in biological neuronal circuits, such as those of the retina. Within the retina, the effect of lateral inhibition is to achieve low-level preprocessing of visual stimuli. One function of such preprocessing is edge enhancement within the primary visual representation. The importance of lateral inhibition in the present context is that it allows the network of nodes to organize into groups, each of which represents one of the data categories, without the aid of feedback from any external agency or teacher. The network is thus capable of self-organization.

The term map in the phrase *self-organizing map* refers to the overall behavior of this network. The network translates, or maps, the data vectors in terms of their category membership onto the array, in this case a two-dimensional array, of network nodes. The result is that the relatively complex space spanned by the training data vectors is mapped onto the relatively simple space of the node array.

As the reference vectors self-organize, the elements of the reference vectors are modified. These modifications are somewhat analogous to the updating of weights in a neural network trained using error backpropagation. Once the process of self-organization has converged (the change per training iteration of the reference vector values has dropped to a low level), the SOM can be used to classify novel exemplars. The SOM now contains a number of groups of reference vectors, each of which represents one of the groups or categories in the training data. The SOM will now assign each novel vector to the category, or group of reference vectors developed during training, which contains the reference vector most closely resembling the novel vector.

Algorithm: Supervised Training—LVQ

In the following description, the terms training vector and test vector refer to the set of predictor values in a training exemplar and test exemplar respectively.

1. Initialize weight vectors $w = [w_1, ..., w_k]$ using pseudo-random values.

2. Assign an equal subset of the weight vectors to each category

3. Define a neighborhood size n, and a value of learning rate r.

4. Repeat the following steps until a criterion has been attained:

    4.1 Training Stage: For each training vector $x = [x_1, ..., x_k]$,

          4.1.1   Find the network node whose weight vector w is nearest to the current training vector x by computing the Euclidean distance d between the training vector and each weight vector w_i:

    $$d = \Sigma_k (x_k - w_{ik})^2$$

          4.1.2   Modify this nearest weight vector according to the rule

              $w_{new} = w_{old} + (r \cdot (x - w_{old}))$ If x belongs to the same category as w;

              $w_{new} = w_{old} - (r \cdot (x - w_{old}))$ otherwise.

    4.2 Testing Stage: For each test vector $y = [y1, ..., yk]$,

4.2.1   Find the network node whose weight vector w is closest to the current test vector y by computing the Euclidean distance d between the test vector and each weight vector wi:

$$d = \Sigma_k (y_k - w_{ik})^2$$

Learning Rate

With the fixed learning rate option, the value of the learning rate is the same for all weight vectors. With the adaptive learning rate option, each weight vector w_j has its own learning rate r_j, defined by the relation

$r_j = r_j / (1 + r_j)$; if the training and weight vectors are in the same category;

$r_j = r_j / (1 - r_j)$; otherwise.

The Learning Vector Quantizer

The LVQ was developed by Teuvo Kohonen as a supervised version of the SOM. Rather than allowing the network nodes to self-organize, the network is guided, using feedback, to develop a set of reference vectors that reflect the categories in the training data. As each training vector is presented to the network, the network reference vector that is most similar to the training vector is identified. Then, that reference vector is modified on the basis of whether the training and reference vectors belong to the same category. If the reference and training vectors belong to the same category, the reference vector is altered to still more closely resemble the training vector. The reference vector is in effect attracted to the training vector. If the two vectors belong to different categories, the reference vector is altered to less closely resemble the training vector. In this event, the reference vector is pushed away from the training vector. Once the LVQ has been trained, then like the SOM, it can be used to classify novel vectors.

Algorithm: Unsupervised Training—SOM

In the following description, the terms training vector and test vector refer to the set of predictor values in a training exemplar and test exemplar respectively.

1.   Initialize weight vectors $w = [w_1, ..., w_k]$ using pseudo-random values.

2.   Define a neighborhood size n, a learning rate value r, and a learning rate decrement factor f.

3.   Repeat the following steps until a criterion has been attained:

3.1 Training Stage: For each training vector $x = [x_1, ..., x_k]$,

      3.1.1   Find the network node whose weight vector **w** is nearest to the current training vector **x** by computing the Euclidean distance d between the training vector and each weight vector w_i:

$$d = \Sigma_k \ (x_k - w_{ik})^2$$

      3.1.2   Modify weight vectors in some neighborhood around the minimum distance node according to the rule

$$w_{new} = w_{old} + (\ g \cdot r \cdot (x - w_{old}))$$

where g is a neighborhood function.

3.2 At some interval, decrease the size of the neighborhood and the value of learning rate according to the rule

$$n_{new} = n_{old} - 1$$

$$r_{new} = r_{old} \cdot f$$

3.3 Testing Stage: For each test vector y = [y1, ..., yk]

      3.3.1   Find the network node whose weight vector **w** is closest to the current test vector y by computing the Euclidean distance d between the test vector and each weight vector wi:

$$d = \Sigma_k \ (y_k - w_{ik})^2$$

Neighborhood Function

The neighborhood function g is defined by the Gaussian function

$$g = exp \ -(\delta \ / \ 2\sigma^2)$$

where δ is the Euclidean distance between the a network node and the center of the neighborhood, and σ is a factor that determines how quickly the weight change decreases with distance from the center. In particular, σ is chosen to be the inverse of the size of the neighborhood. If the size of the neighborhood is set to 0 then g = 1.

    Unlike other network functions in Simulnet, the *Vector Quantizer Network* function is specifically designed to be used for classification. It is not designed to operate as a universal function approximator. However, under the right conditions, the *Vector Quantizer Network* function is capable of operating as an optimum Bayesian classifier, performing pattern classification with the minimum possible rate of misclassification. As with other network classifiers, these conditions include the primary one, that there is a sufficient number of training vectors and that the classes being discriminated are well represented in the training data.

The Simulnet Vector Quantizer Network Function

In this section the various controls and settings that are available with the *Vector Quantizer Network* function are listed and described. The *Vector Quantizer Network* function implements two network models: an unsupervised-training, self-organizing, feature map and a supervised-training, learning vector quantizer. Table 2.4 lists and describes the available controls and settings located on the *Vector Quantizer Network* dialog form. This table does not include some of the functions that have already been discussed in the section dealing with the *BackProp Network* function.

Before attempting to use the *Vector Quantizer Network* function, the reader is strongly encouraged to run Simulnet, select the *Vector Quantizer* function from the *Network* menu to show the *Vector Quantizer Network* dialog form, and then study the information in Table 2.4. Going through this exercise is particularly important if the reader wishes to use this function with data other than that which has been supplied with Simulnet.

Table 2.4 *Vector Quantizer Network* **Controls**

Controls Located on *Vector Quantizer Network* dialog form

Control	Description
Training Mode	
Supervised vs. Unsupervised	Two training mode options are available. The Supervised (LVQ) option allows the network to be trained with feedback using the LVQ algorithm. The Unsupervised (SOM) option allows training to be carried out without feedback using the SOM algorithm.
Network Architecture	
Number of network nodes	The network consists of a rectangular array of nodes, with the size of the array specified by the *No. of rows* and *No. of columns* parameters.
Network Archiving	
Saving a network	A network configuration can be saved to disk at any time, using the *Save Net* button.
	A network configuration consists of the values of the network weights, with a header specifying the network configuration which contains the number of network rows and columns.
Recreating a network	A network that has been saved as a disk file can later be recreated using the *Load Network: File* button. Using these functions, network training can be stopped at any point and

the network saved to disk. The network can then be recreated at any future time and training resumed.

This feature allows network weights to be initialized from a file rather than by means of a pseudo-random number generator.

Display Objects

Status

The *Display: Status* option shows the number of the current training iteration. The *Training: Correct* indication on this display represents the percentage of training exemplars that have been correctly classified. The *Testing: Correct* indication on this display represents the percentage of testing exemplars that have been correctly classified. These figures are updated every training iteration.

Error History

The *Display: Error History* option displays a graph of training performance (percentage of training exemplars correctly classified) and testing performance (percentage of testing exemplars correctly classified) from the start of training. The number of iterations per graph is specified by the value of the *Passes* parameter. When this number of iterations has been graphed, the graph is cleared and restarted beginning with the next iteration.

Data Matrix

The *Display: Data Matrix* option displays a matrix of training and testing target values and output-node values. Data are arranged in four columns, with one row for each training or testing exemplar. Column 1 holds the target outcome values for each training exemplar. Column 2 holds the corresponding outcome values predicted by the network. Column 3 holds the target outcome values for each testing exemplar. Column 4 holds the corresponding value predicted by the network.

Nodes

The *Display: Nodes* option displays a color-coded representation of the category membership of each network node. In the *Supervised (LVQ)* training mode this display is static: Network nodes are each assigned to one of the data categories at the time that the network is started, or restarted, and this category does not change during training. In the *Unsupervised (SOM)* training mode, this display will show the ongoing changes in category membership of each node, updated every training iteration.

Training Parameters

Learning Rate	Learning rate can be preset at fixed value (the *Set* option), or allowed to vary with a unique value for each network node (the *Adaptive* option). Training error convergence with many types of training data should generally be faster with the adaptable learning rate option.
	To use fixed learning rate, select the *Training: Learning rate: Set* option. The same value of learning rate is used for all nodes. This value is entered in the associated data entry field.
	To use variable learning rate, select the *Training: Learning rate: Adaptive* option. A unique learning rate value is computed for each node. This value is updated in each training iteration based on that node's on-going training performance. The initial value of learning rate is the value entered in the *Training: Learning rate: Set* data entry field.
Neighborhood	This parameter specifies the size of the sensitive region around each network node. The values of network weights associated with nodes that fall within this sensitive region will be updated in every training iteration. The shape of the neighborhood is a square.
Update Interval	This option specifies how frequently the size of the neighborhood and the value of learning rate are reduced. Updates are carried out once every *Update Interval* training iterations. For example, if *Update Interval* is set to 10, then neighborhood size and learning rate are adjusted every 10 training iterations. Neighborhood size is reduced by one during each update. Learning rate is multiplied by a factor of 0.8 every update. A value of 0 disables updates.

Simulnet Exercise: Predicting Chaotic Data

In this exercise the *Vector Quantizer Network* will be asked to classify data vectors. The data vectors will be generated by the cubic map. The cubic map is a system of two coupled difference equations,

$$x(i + 1) = y(i)$$

$$y(i + 1) = -a\, x(i) + b\, y(i) - y(i)^3$$

The cubic map can exhibit a variety of behaviors depending on the values of parameters a and b. In particular, the map will behave chaotically for parameter values in the neighborhood of a = 0.2 and b = 2.77. To create the training and testing data, the cubic map was iterated using these parameter values, generating a correspondingly random-looking sequence of data points (Figure 2.19). This sequence was divided into triplets of data points, each triplet composed of three consecutive points in the sequence. The *Vector Quantizer Network* will be asked to make the following prediction: Given the first two data points in a triplet, will the third point in the triplet be greater or smaller than the second point in the triplet? Actually, the value generated by the cubic map depends only on the single preceding value generated by the map. Only one data value is therefore actually required as the predictor. Nevertheless, in this exercise two preceding values are actually used to train the network.

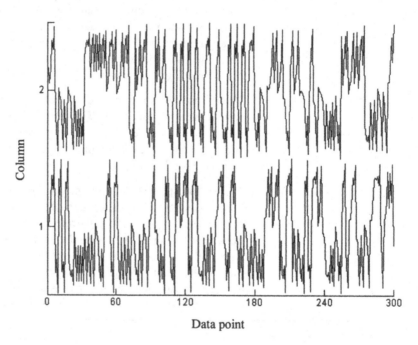

Figure 2.19. The figure shows each of the two variables of the cubic map; x in the lower graph and y in the upper graph. To produce this data the cubic map was iterated 300 times.

In more detail, the following procedure was used to generate the network training and testing data. For each data set, the cubic map was iterated 300 times using a random starting value generating a sequence of 300 data points. These 300 data points were then reorganized into 100 triplets of points. These 100 triplets were organized as a matrix with 3 columns of 100 rows. In each row, columns 1 and 2 contain the first and second data points of each triplet. The values in column 3 are either a 0 or a 1 according to the following rule: If the third data point in a triplet is more positive than the second data point of the triplet, the value in column 3 is a 1; otherwise, the value in column 3 is a 0.

In terms of the original chaotic time series, within each matrix row, the data points in successive columns follow each other in time:

$$
\begin{array}{l}
\textit{exemplar 1} \\
\textit{exemplar 2} \\
\cdot \\
\cdot \\
\cdot \\
\textit{exemplar 100}
\end{array}
\left[
\begin{array}{lll}
\text{time-point 1} & \text{time-point 2} & (0 \text{ or } 1) \\
\text{time-point 4} & \text{time-point 5} & (0 \text{ or } 1) \\
\cdot & \cdot & \cdot \\
\cdot & \cdot & \cdot \\
\cdot & \cdot & \cdot \\
\text{time-point 298} & \text{time-point 299} & (0 \text{ or } 1)
\end{array}
\right]
$$

Each row of this matrix constitutes 1 network exemplar: For each exemplar there are 2 continuous independent variables in columns 1 and 2, and 1 categorical dependent variable in column 3. The network will be asked to predict the value in column 3; whether the next value to be generated by the cubic map, behaving chaotically, will be greater or less than the previous value, given the preceding two values. In other words, the network will be asked to classify each exemplar according to whether the exemplar is followed by an increasing or a decreasing trend.

Two sets of data points were generated using this procedure. The training set is stored in file *cubic2.trn*, and the testing set is stored in file *cubic2.tst*. Each set was constructed to be independent of the other by initializing the cubic map each time with a new, random value. The network will be trained on the training set, and tested (or validated), using the testing set. As the network is being trained, validation (or testing), will be periodically carried out by applying the testing data once every five training iterations (passes through the training data).

Procedure

1. Close all forms on the desktop. From the *Network* menu, select the *Vector Quantizer Network* option. On the *Vector Quantizer Network* form, click the *Load Setup* button. On the *Load Setup* dialog form, select file *cubic2ln.set*. This will load training file *cubic2.trn*, and testing file *cubic2.tst*. A number of other setup parameters will be loaded as well, in particular, a value of smoothing of 0.05.

2. On the *Vector Quantizer Network* dialog form, click the *Train* button. A number of objects will appear on the desktop: A status form showing the current network training and testing errors and the pass number; a matrix form that will be used to hold the network results; and a graph form on which will be plotted a record of the network training and testing classification rate, averaged over the corresponding exemplars in each data set. The network will then start training.

   As the network trains, training and testing classification rates should rise to values in the region of 85 to 95%. These figures are the percentages of the exemplars in the corresponding data sets that have been correctly classified. Correct classification, in this case, means that the *Vector Quantizer Network* has correctly predicted whether the next data value will be higher or lower than the preceding data value.

The number of passes needed to train the network to criterion cannot be determined precisely, but it will probably take between 100 and 200 passes.

When the *Training: percent correct* value has reached 100%, network training will stop. Alternatively, training can be halted after *Training: percent correct* has reached some level such as 90 or 95%. At this point, the results matrix can be examined. There are 4 columns in this matrix: Columns 1 and 2 hold the target and network scores for the training exemplars. Columns 3 and 4 hold the target and training scores for the testing exemplars. Columns 3 and 4 can be plotted to show how accurately the network was able to learn the chaotic time series. This plot is shown in Figure 2.20.

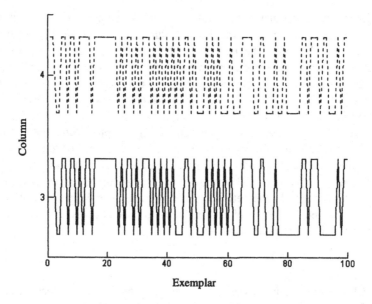

Figure 2.20. The bottom (solid) graph of column 3 of the network results matrix shows the target values that the network was asked to predict. Though no apparent regularity exists in the time series generated by the cubic map, the network was able to learn to predict these values, as shown in the top (dotted) plot of column 4 of the network results matrix.

The network results matrix consists of 4 columns and 100 rows. Each row represents one exemplar. Columns 1 and 2 relate to the training exemplars. Column 1 shows the dummy code that was assigned to each exemplar in the training file when the file was created. Column 2 of the network results matrix shows the corresponding value as predicted by the network. Columns 3 and 4 relate to the testing exemplars. Column 3 shows the target value from each testing exemplar. The network does not use this target value in any way. It is presented in the network results matrix solely for the convenience of the network user so that the performance of the network can be assessed. Column 4 shows the corresponding value that the network predicted for each of the testing exemplars. The degree of match between the values in columns 3 and 4 indicates the extent to which the network has learned the assigned task; predicting the direction of change of the next data point, based on the values of the previous two points.

References

Kohonen, T. (1989). *Self-Organization and Associative Memory (3rd ed.)*. Berlin: Springer-Verlag.

Assessing the Significance of Network Results

Introduction

The statistical significance of the results of an analytical procedure, such as neural network classification, can be assessed by applying an appropriate statistical test. The particular test that is used depends on the nature of the results, that is, the type of analysis that the network has carried out, as well as on the question that is to be answered regarding the results.

In a typical application an experiment would be conducted involving some sort of manipulation that differentiates between, say, two conditions. Data would be recorded within each of these conditions. The experiment could, of course, consist of more than two conditions. In any event, the question to be answered is, has the experimental manipulation resulted in differences between the data measured in the two conditions?

This question is addressed as follows: Using the data for each of the conditions, a set of network exemplars is created. This produces two groups of exemplars, one for each of the two conditions in the experiment. Next, the exemplars are divided into two groups, a training set and a test set. Let us suppose that there are available a total of 100 exemplars for condition 1 of the experiment, and 100 exemplars for condition 2. Of these, 60 condition 1 exemplars, selected at random from the pool of 100, and 60 condition 2 exemplars, also randomly selected, could be used to form the training set. The balance, 40 from conditions 1 and 2, could then be used to form the test set.

A neural network could then be trained to distinguish between the two groups of exemplars in the training set. Once the network has been trained, it could be used to classify the exemplars in the test set. The result would be a set of scores, one for each exemplar in the test set. Because the experimental condition to which each test exemplar belongs is known, the scores for the test exemplars can be separated into two groups, one for each condition.

The original question can now be addressed by seeing if there is a difference between the scores for the condition 1 test exemplars and the condition 2 test exemplars. To assess the significance of this difference an appropriate statistical test would be applied to the scores. If a statistically significant difference can be shown between the two groups of scores, then, with some measurable degree of confidence, it could be concluded that the network was able to distinguish between the two groups of test exemplars. Substantively, that is in terms of the substance of the experiment itself, it could be concluded that there is support for the hypothesis that the experimental manipulation that generated the test exemplars has had a significant effect in the experiment.

When a neural network is used to classify test exemplars into two categories, one of the simplest statistical tests to apply and interpret is the Student's t-test. This test addresses the following question: What is the probability that the observed difference between the means of the two groups of network scores is due to chance? The smaller this probability, the more likely it is that the difference in means reflects a real difference between the two groups of scores. In terms of the extent to which the neural network was able to learn to classify the novel exemplars, a small value of probability associated with the t-test suggests that there was a correspondingly small probability that the classification performance of the network was due to chance.

Simulnet Exercise: The Student's t-test

The value of Student's t-test indicates how large the difference between two group means is, relative to the variance of the two groups. An associated value of probability then indicates how likely it is that a difference in group means could have been obtained by chance. A typical application for this statistical test is to estimate the extent to which an experimental manipulation has been successful in creating a difference in outcome between two groups. In this exercise the t-test will be applied to a set of network classification scores.

The network scores are contained in a matrix of two columns and some number of rows. Each row of the network results matrix represents one exemplar in the training file. Column 1 shows the dummy code that was assigned to each exemplar in the training file when the exemplar was created. Some exemplars are coded with a zero, indicating that they belong to the first experimental condition. The remaining exemplars are coded with a one, indicating that they belong to the second experimental condition. These dummy codes are shown in this network results matrix for the convenience of the network user only. The network does not use these codes in any way. Column 2 of the network results matrix shows the actual score assigned to each of the exemplars by the network.

Procedure

1. Close all forms on the desktop. Open file *class1.dat* and minimize the matrix form. This file contains the results from a network training session.

2. From the *Analyze* menu, select the *Statistics - Inferential* option. On the *Inferential Statistics* dialog form, click on the *Type of analysis* list-box to show the list of available options. Scroll through the list to find the *t-test (unequal variance)* function. Select the *Output: text* option. For *Data group 1: starting column* enter a value of 2. For *Data group 1: ending column* enter a value of 2. For *Data group 2: starting col-*

umn enter a value of two. For *Data group 2*: *ending column* enter a value of 2. For *Data group 1*: *starting row* enter a value of 1. For *Data group 1*: *ending row* enter a value of 5. For *Data group 2*: *starting row* enter a value of 6. For *Data group 2*: *ending row* enter a value of 10.

3. Click the *Compute* button. A text-box containing the results of the analysis will appear, as shown below.

T-Test of Distributions with Unequal Variances

Item	Group 1	Group 2
Starting row	1	6
Ending row	5	10
Starting column	2	2
Ending column	2	2
Number of cases	5	5
Mean	0.03	0.876
Variance	0.00294	0.065
df	4.36	
Pooled variance	0.014	
t-value	-7.249	
probability:		
one-tailed	0.000685	
two-tailed	0.00137	

The critical value in this set of data is *probability: one-tailed*. This value indicates the significance of the difference between the group means. If this value is less than or equal to 0.05, it may be stated that there appears to be a significant difference between the two groups of scores. If the value is between 0.1 and 0.05, the difference in means may be referred to as being marginally significant. If the value is greater than 0.1, then all that can be said is that a significant difference has not been found.

With a statistical test, absence of evidence is almost never evidence of absence. If the test fails to find a significant effect in the data, it cannot be claimed that the data does not contain a significant difference. All that can be said is that a difference has not been found. It may be that the data contains a significant effect, but that the noise level in the data is too high to allow the effect to be detected with the statistical test. In the case of the t-test, a nonsignificant value of t may only indicate that there was a real difference in the means of the two groups of scores, but that the within-group variance of one or both groups was too large to allow the difference in means to be detected.

Network Application Examples

Introduction

This section describes a number of network application examples that have been provided to allow the reader to become acquainted with the operation of the network functions in Simulnet. Each example describes how each of the neural network functions can be applied to a particular type of problem. The examples are intended to be, to some extent, self-contained, and assume only that the reader has become familiar with the Simulnet interface. Inevitably then there is some duplication of information between examples. The first example does, however, contain a more detailed description of the procedures to be followed in using the networks. For this reason it is suggested that the reader work through Example 1 before proceeding to any of the other examples.

The examples are intended to demonstrate how the network functions may be used to solve classification, function learning, and prediction problems. Each example uses a set of three disk files that have been provided.

- A setup file: A file containing the values of a set of operating parameters for a particular network function. Setup file names have the extension .set.
- A training set: A file containing a set of training exemplars. Training file names have the extension .trn.
- A testing set: A file containing a set of testing exemplars. Testing file names have the extension *.tst*.

Training and testing files are organized as matrices. Each matrix row contains one exemplar of the relationship to be learned. Each matrix column contains the values of the independent and dependent variables of this relationship.

Setup files are text files containing the values and settings for all parameters on a network dialog form. A configuration of settings on a network dialog form can be saved as a setup file at any time.

Running the Examples

To run the following examples, select a network by clicking on one of the network toolbar buttons or select one of the network options from the *Network* menu. When the network dialog form appears, click the *Load Setup* button. On the *Load Setup* dialog form select one of the files with a *.set* extension, as indicated in the following examples. All files will be located, by default, in the directory *\Simulnet\Data*. Setup files are named according to the network function they are intended to be used with.

Network	Setup File Name
BackProp Network	xxxxnn.set
Genetic Network	xxxxgn.set
Probabilistic Network	xxxxpn.set
LVQ Network	xxxxln.set,
	xxxxsn.set

The first part of the setup file name (shown above as xxxx) will refer to the contents of the particular file. For example, file xornn.set contains the *BackProp Network* setup for the XOR function. Parameters contained in the setup files have been prepared to provide demonstrations of capabilities of the various network functions. These parameters may not be optimal in terms of, for example, network size or training time. The user is therefore encouraged to experiment with variations on the supplied settings. To a great extent network behavior is problem dependent. In order to develop some level of intuition about what can be expected from each of the network functions, the user should be prepared to devote time to observing what a network function does as a function of changes in the values of its operating parameters. These observations should first be made using the supplied setup and data files, and later on the user's own data. If a supplied setup is changed, the altered setup should be saved using a name different from that of the original setup file.

Network Information Displays

In order to demonstrate the use of the various network information displays, the provided setup files have been designed to invoke these displays whenever appropriate. Using these displays, however, will slow the rate at which the networks train because a certain amount of time is needed to update the displays once every training iteration. There are two components to the time needed to complete a training iteration: the time needed to update information displays and the time needed to carry out the training computations. The amount of time required to update the information displays depends on the size of the network: the number of input, hidden, and output nodes. The larger the network, the more time needed to update most of the displays. The amount of time required to carry out the network training computations depends on the size of the training file: The larger the training file, the more time needed to carry out the computations. For any given network size, as the size of the network training file increases, the proportion of time needed to update the displays will decrease. The largest time penalty will occur when a moderate or large network is trained using a small training file. This situation may arise with some of the supplied network setups. With real-world data however, involving moderate to large training files, the extra time needed to update the information displays should be a reasonably small proportion of the time needed to complete a training iteration.

When running the provided setups, the user may want to disable some of the network information displays once it has become clear what it is that is being presented. An information display can be disabled by minimizing the associated display form. A minimized display form will not be updated. To resume updating of the display, restore the display form to normal size. Information displays can be kept minimized during a training session, and restored to normal size periodically to inspect network progress.

Example 1: Sine-Cosine Discrimination

In this example, the networks are required to learn to distinguish between noisy sine and cosine waveforms. The training sets each contain 10 exemplars: 5 of a sine wave with Gaussian noise added, and 5 of a cosine wave with Gaussian noise added. Each exemplar consists of 24 predictor values that define a noisy sine or cosine wave, along with 1 criterion value, a 0 or 1. A criterion value of zero labels the exemplar as a sine wave. A criterion value of one labels the exemplar as a cosine wave. The testing set is organized in exactly the same way. Although the exemplars in the test file contain criterion values, the networks do not use these values. Criterion values in the testing set are provided for reference only, in order that the user can evaluate the performance of the network.

In the training phase of the *BackProp Network* and *Genetic Network*, the networks are repeatedly presented with the exemplars in the training set. Through this process, the networks learn the relationship between the predictor variables (the numerical values defining the sine and cosine waves) and the criterion variables (the dummy code defining the category of the waveform). Recall that for the *Probabilistic Network* there is no analogous training phase; the training data is simply loaded from the disk file into working memory. In the test phase, and having learned the relationship between predictor and criterion values, the networks are asked to predict the category of the 40 exemplars in the test set.

Select the network function by clicking the appropriate toolbar button, or by choosing an option from the *Network* menu. Then click the *Load Setup* button on a network dialog form to select the appropriate setup file from the following list:

Setup Files for Sine-Cosine Discrimination Example

Network Function	Setup File
BackProp Network	sincosnn.set
Genetic Network	sincosgn.set
Probabilistic Network	sincospn.set
Self-organizing Map	sincossn.set
Vector Quantizer Network	sincosln.set

When the setup file has finished loading, the network is ready to be trained. To begin training use the procedure from the following sections that applies to the particular network being used.

BackProp Network, Genetic Network, Vector Quantizer Network

Click the *Train* button. A number of objects will be loaded onto the desktop, and training iterations will begin. When the specified stopping criterion has been achieved, training will stop. Network training and testing results are contained in the network results matrix. Each row in this matrix contains values for one exemplar. The first two columns in this matrix show the target and computed scores for the exemplars in the training set. The last two columns show the target and computed scores for the exemplars in the testing set.

The extent to which the networks have learned the required discrimination task is indicated by the difference between the values in column 1, target scores for training exemplars, and column 2, predicted scores for training exemplars. If the differences between the target and predicted scores are small, it could be concluded that the network has learned to distinguish between the sine and cosine exemplars. The extent to which the network can generalize this training is demonstrated by the difference between the values in column 3, target scores for test exemplars, and column 4, predicted scores for test exemplars. Again, if the differences between the target and predicted scores are small, it would indicate that the network is able to generalize what it has learned to the testing exemplars.

Training using the *Genetic Network* will generally take fewer passes than using the *BackProp Network*, demonstrating the *Genetic Network*'s relative immunity to entrapment by local features of the problem search space. However, the increased computational requirements of the *Genetic Network* generally mean that this network will take more time to converge than the *BackProp Network*. On a classification task of this sort, the *Vector Quantizer Network* will generally take fewer iterations to converge to a low value of error then either the *BackProp* or *Genetic* networks.

For the *BackProp Network* and the *Genetic Network* network, performance is indicated by the values on the *Status* form. These values indicate the RMS and average errors, computed over all exemplars and all network outputs, and the percentage of exemplars for which outputs were within the specified distance from the target values.

Restarting Network Training

When training the *BackProp Network*, it is possible for the network to locate, and become trapped in, a region of the search space corresponding to a good, but not optimal, solution. Alternatively, the network may enter an extended flat region of the problem search space. If either situation occurs the neural network can take hundreds or even thousands of iterations to converge to a low error level. While this will probably not happen in the current example, it may occur in some of the more difficult tasks that are involved in the subsequent examples. If the network does not appear to be making progress, indicated by relatively constant values of training error on the network *Status* display, a useful strategy is simply to randomize network weights and restart training. Follow these steps to restart training.

1.   Stop network training by clicking on the *Stop Process* toolbar button .

2. Click the *Restart Network* toolbar button [image]. Click 'Yes' on the message box that will appear.

3. Finally, click the *Train* toolbar button [image]. Network training will resume using a new, randomized set of weight values.

Probabilistic Network

Click the *Compute* button on the network dialog form. The *Probabilistic Network* uses a single-pass algorithm, and computation should take only a few seconds to complete. Results are contained in the network results matrix that will appear on the desktop. Each row in this matrix contains values for one exemplar. The first column in this matrix shows the target scores for each exemplar in the testing set. The second column shows the predicted scores for each exemplar in the testing set. Because there is no actual training phase with this network, there are no columns in this matrix for the training set. The extent to which the network is able to perform the classification task is indicated by the difference between the target values in column 1 and computed values in column 2 for each testing exemplar. These differences can be seen to be reasonably small, indicating that the network is able to accurately classify the sine and cosine exemplars.

The general network training procedures that have been outlined in this example apply to the rest of the examples that are described in this section. For this reason the following examples contain relatively abbreviated descriptions of the procedures.

Example 2: Learning the XOR Function

In this example the networks are required to learn the exclusive-OR (XOR) function. The XOR function takes two input values, and generates one output value. The behavior of the function is summarized in the truth table for the function, shown in the following table. A truth table is simply a description of the inputs and corresponding outputs for a function that operates on binary, that is 0 and 1, values.

Exclusive-Or Truth Table

Input 1	Input 2	Output
0	0	0
0	1	1
1	0	1
1	1	0

Straightforwardly, the XOR function generates an output of one if either of the two inputs is one, but not if both inputs are one. In order to learn this function, the network must develop a nonlinear decision surface. In other words, there is no way in which the two categories represented by the zero and one output values can be separated by a flat surface in the space defined by the two inputs. For this reason the XOR problem is an example of a relatively difficult neural network classification problem.

The training data contains 24 exemplars of the XOR function. Each exemplar consists of two predictor values that define the inputs to the XOR function, along with one criterion value that defines the output of the XOR function. The test data is organized in the same way, and contains four exemplars of the XOR function.

The *Vector Quantizer Network* treats the XOR task as a classification problem, classifying exemplars into category 0, corresponding to the XOR output of 0, or into category 1, corresponding to the XOR output of 1. On problems that can be cast into the form of classification tasks, the *Vector Quantizer Network* should generally converge to a low-error level more quickly then either the *BackProp* or *Genetic* networks.

Click the *Load Setup* button on a network dialog form to select the appropriate file from the following list:

Setup Files for XOR Learning Example

Network Function	Setup File
BackProp Network	xornn.set
Genetic Network	xorgn.set
Probabilistic Network	xorpn.set
Self-organizing Map	xorsn.set
Vector Quantizer network	xorln.set

Example 3: Learning the Odd-Parity Function

Like the previous example, this example also involves a function learning task, learning a form of the odd-parity function. The odd-parity function is a generalization of the exclusive-or function. The output of the function is one if the number of inputs that are one is odd, and the output is zero otherwise. The truth table for this function, for the case of three inputs, is shown in the following table.

Odd-Parity Function Truth Table

Input 1	Input 2	Input 3	Output
0	0	0	0
0	0	1	1
0	1	0	1
0	1	1	0
1	0	0	1
1	0	1	0
1	1	0	0
1	1	1	1

As for the XOR function, in order to learn the odd-parity function the network must also develop a nonlinear decision surface. There is no way in which the two categories represented by the zero and one output values can be separated by a flat surface in the space defined by the two inputs. Learning the odd-parity function is actually a more difficult neural network learning task than the XOR function.

The training set consists of eight exemplars, corresponding to the eight rows of the truth table. Each exemplar consists of three predictor values that define the inputs to the odd-parity function, along with one criterion value that defines the output of the function. The testing set is identical to the training set.

As in the case of the XOR function, the *Vector Quantizer Network* treats the odd-parity learning task as a classification problem, classifying exemplars into category 0, corresponding to the odd-parity output of 0, or into category 1, corresponding to the odd-parity output of 1.

Click the *Load Setup* button on a network dialog form to select the appropriate file from the following list:

Setup Files for Odd-parity Learning Example

Network Function	Setup File
BackProp Network	oddparnn.set
Genetic Network	oddpargn.set
Probabilistic Network	oddparpn.set
Self-organizing Map	oddparsn.set
Vector Quantizer network	oddparln.set

Example 4: Learning Phase Relationships

In this example, the network will be required to learn a functional relationship between a set of independent and dependent variables. This learning task can be viewed as a prediction problem; predicting the value of the dependent variables, given the values of the independent variables. In a prediction task, the networks are given the opportunity to learn a functional relationship by being presented with examples of that relationship.

The functional relationship that the networks are required to learn in this example is the relative phase angle of a one-cycle sine wave with added Gaussian noise. The training set consists of 40 exemplars, organized into 4 equal-sized groups. The exemplars in the first group contain sine waves with a phase angle of 0 degrees. The exemplars in the second, third, and fourth groups contain sine waves with phase angles of 30, 60, and 90 degrees respectively. Each exemplar consists of 24 predictor values that define the noisy sine wave, along with 1 criterion value. This criterion is the numerical value of phase angle of the sine wave: 0, 30, 60, or 90. The test set consists of seven exemplars, with each exemplar organized in the same way as the exemplars in the training set. The 7 test exemplars represent noisy sine waves with phase angles of 0, 15, 30, 45, 60, 75, and 90 degrees respectively.

Click the *Load Setup* button on a network dialog form to select the appropriate file from the following list:

Setup Files for Phase Learning Example

Network Function	Setup File
BackProp Network	phasenn.set
Genetic Network	phasegn.set
Probabilistic Network	phasepn.set
Self-organizing Map	phasesn.set
Vector Quantizer Network	phaseln.set

Example 5: Predicting the Logistic Map 1

This example also involves a prediction task. The networks are again required to learn a functional relationship by being trained on multiple exemplars of that relationship, and to then predict the value of a dependent (criterion) variable, given the value of an independent (predictor) variable. The function that the networks are required to learn is the relationship between a single datum in a chaotic time series, and the immediately following point in the series. The chaotic time series was generated by iterating the logistic map:

$$x(n + 1) = x(n) + G\, x(n)\, (1 - x(n))$$

where the parameter G controls the behavior of the equation. For a value of G = 3, the equation behaves chaotically. To generate the data for this example, the logistic map was iterated 200 times with G = 3 and, starting from a randomly selected initial value, creating a time series of 200 data points. This time series was transformed using the *Matrix Tools* function *Interleave Rows*, using a value of two rows. This transformation converts the original 200 point series into a matrix of 2 columns by 100 rows, with each row of the matrix containing in its 2 columns, 2 consecutive points from the original series. This matrix becomes the training set for the network. Column 1 contains the independent (predictor) variable, while column 2 contains the dependent (criterion) variable. In this way, the network will be presented repeatedly with a single point from the time series as the predictor, and the next point in the series as the criterion value. The network will thus be required to learn to predict the value of the point immediately following any point on the chaotic time series. The training data is stored in file *log.trn*. The testing data set was constructed in the same way, by iterating the logistic map for 200 points, and starting from a new, randomly selected initial value. This testing data is stored in file *log.tst*.

Click the *Load Setup* button on a network dialog form to select the appropriate file from the following list:

Setup Files for Chaotic Data Prediction Example

Network Function	Setup File
BackProp Network	lognn.set
Genetic Network	loggn.set
Probabilistic Network	logpn.set

Example 6: Predicting the Logistic Map 2

This example again involves a prediction task. As in the previous example, the networks are required to learn the logistic function. In this example, however, the networks are required to learn the relationship between two consecutive points generated by the logistic map, and the immediately following point. Starting from a random initial value, the logistic map was iterated 300 times, producing a time series of 300 data points. This time series was transformed using the *Matrix Tools* function *Interleave Rows*, using a value of three rows. This transformation converted the original 300 point series into a matrix of 3 columns by 100 rows. Each row of the matrix contains, in its three columns, three consecutive points from the original series.

This matrix becomes the training set for the network. Columns 1 and 2 contain the predictor variables while column 3 contains the criterion variable. Using this training data, the network will be presented repeatedly with two consecutive points from the chaotic time series as predictor values, and the immediately following point as the criterion value. The network will thus be required to learn to predict the value that would be generated by the logistic map given two consecutive values. This training data is stored in file *log1.trn*. A testing data set was constructed in the same way, by iterating the logistic map for 300 points, and starting from a new, random initial value. This testing data is stored in file *log1.tst*.

Click the *Load Setup* button on a network dialog form to select the appropriate file from the following list:

Setup Files for Chaotic Data Prediction Example

Network Function	Setup File
BackProp Network	log1nn.set
Genetic Network	log1gn.set
Probabilistic Network	log1pn.set

Example 7: Predicting the Logistic Map 3

This example involves a classification task. In the previous example, the networks were required to learn the logistic map function. In the present example, the networks will instead be asked to predict whether the value generated by the logistic map will increase or decrease, given the preceding two values.

As in the previous example, the data used to train the network was generated by iterating the logistic equation. Starting from a randomly selected initial value, the logistic map was iterated 300 times, producing a time series of 300 data points. This time series was transformed using the *Matrix Tools* function *Interleave Rows*, using a value of three rows.

This transformation converts the original 300 point series into a matrix of 3 columns by 100 rows, with each row of the matrix containing in its 3 columns, 3 consecutive points from the original series. The value in this third column was then replaced with either a 0 or a 1 according to the following rule: If the value in column 3 was more positive than the value in column 2, the value in column 3 was replaced with a 1; if the value in column 3 was less positive than the value in column 2, the value in column 3 was replaced with a 0.

This matrix becomes the training set for the network. Columns 1 and 2 contain the predictor variables while column 3 contains the criterion variable. The network will be presented with exemplars that consist of two data points from the chaotic time series as predictor values, and dummy codes indicating the trend of those points, rising or falling, as criterion values. The network will thus be required to learn to predict, given two consecutive data points, whether the data point that follows will be higher or lower in value than the second of the two data points. This training data is stored in file *log1.trn*. A testing data set was constructed in the same way, by iterating the logistic map for 300 points and starting from a new, randomly selected, initial value. This testing data is stored in file *log1.tst*.

Click the *Load Setup* button on a network dialog form to select the appropriate file from the following list:

Setup Files for Chaotic Trend Prediction Example

Network Function	Setup File
BackProp Network	log2nn.set
Genetic Network	log2gn.set
Probabilistic Network	log2pn.set
Self-organizing Map	log2sn.set
Vector Quantizer network	log2ln.set

Example 8: Predicting the Cubic Map 1

In this example, the function that the networks are required to learn is the cubic map. The cubic map is a system of two coupled difference equations:

$$x(i + 1) = y(i)$$

$$y(i + 1) = -a\, x(i) + b\, y(i) - y(i)^3$$

The cubic map was suggested by Holmes (1979) as an approximation to a Duffing oscillator. A strange attractor exists for parameter values in the neighborhood of a = 0.2 and b = 2.77. Using these parameter values, and starting from a randomly selected initial value, the cubic map was iterated 300 times, producing a time series of 300 data points. This time series was transformed using the *Matrix Tools* function, *Interleave Rows*, using a value of three rows. This transformation converted the original 300 point series into a matrix of 3 columns by 100 rows, with each row of the matrix containing 3 consecutive points from the original series. This matrix is stored on disk as file *cubic1.trn*, and will be the training matrix for the network. Columns 1 and 2 contain the independent (predictor) variables, while column 3 contains the dependent (criterion) variable. In this way, the network will be presented repeatedly with two points from the chaotic time series as predictor values, and the immediately following point as the criterion value. The network will thus be required to learn to predict the value of the point immediately following any pair of consecutive predictor points. A testing data set was constructed in the same way, by iterating the cubic map for 300 points and starting from a new, randomly selected initial value. This testing data is stored in file *cubic1.tst*.

Click the *Load Setup* button on a network dialog form to select the appropriate file from the following list:

Setup Files for Chaotic Data Prediction Example

Network Function	Setup File
BackProp Network	cubic1nn.set
Genetic Network	cubic1gn.set
Probabilistic Network	cubic1pn.set

Reference

Holmes, P. J. (1979). *A nonlinear oscillator with a strange attractor*. Philosophical Transactions of the Royal Society of London, A, 292: 419-448.

Example 9: Predicting the Cubic Map 2

This example involves a classification task. In the previous example, the networks were required to learn the principle underlying the chaotic time series data generated by the cubic map. In the present example, however, the networks will be asked to predict whether the value generated by the cubic map will be larger or smaller, given the immediately preceding two values. The data used to train the network was generated by iterating the cubic map, using the same parameter values as in the previous example. Starting from a randomly selected initial value, the cubic map was iterated 300 times, producing a time series of 300 data points. This time series was again transformed using the *Matrix*

Tools function, *Interleave Rows*, using a value of three rows. This transformation converts the original 300 point series into a matrix of 3 columns by 100 rows, with each row of the matrix containing in its 3 columns, 3 consecutive points from the original series. The value in this third column was then replaced with either a 0 or a 1 according to the following rule: If the value in column 3 was more positive than the value in column 2, the value in column 3 was replaced with a 1; if the value in column 3 was less positive than the value in column 2, the value in column 3 was replaced with a 0.

This matrix has been saved as file *cubic2.trn*, and will be the training set for the network. Columns 1 and 2 contain the predictor variables while column 3 contains the criterion variable. In this way, the network will be presented repeatedly with exemplars that consist of two data points from the chaotic time series as predictors, and dummy codes indicating the trend of those points, rising or falling, as criterion values. The network will thus be required to learn to predict, given two consecutive data points, whether the data point that follows will be higher or lower in value than the second of the two data points. A testing data set was constructed in the same way, by iterating the cubic map for 300 points and starting from a new, randomly selected initial value. This testing data is stored in file *cubic2.tst*.

Click the *Load Setup* button on a network dialog form to select the appropriate file from the following list:

Setup Files for Chaotic Trend Prediction Example

Network Function	Setup File
BackProp Network	cubic2nn.set
Genetic Network	cubic2gn.set
Probabilistic Network	cubic2pn.set
Self-organizing Map	cubic2sn.set
Vector Quantizer network	cubic2ln.set

Example 10: Predicting the Lorenz Flow

This example involves a prediction task in which the networks are required to learn the Lorenz flow, a system of three coupled ordinary nonlinear differential equations:

$$x' = \sigma (y - x)$$

$$y' = Rx - y - xz$$

$$z' = xy - Bz$$

The Lorenz flow was suggested by Edward Lorenz (Lorenz, 1963) as a simplified model of convection dynamics. This system exhibits chaotic behavior for values of the 3 control parameters of $\sigma = 10$, $R = 28$, and $B = 8/3$. Using these parameter values and starting from a set of random initial values, the Lorenz flow was iterated 1,000 times producing a matrix of 3 columns, 1 for each system variable, of 1,000 data points each.

To create the network training matrix this time series was transformed using the *Matrix Tools* function, *Stack Rows*, using a value of two rows. This transformation converted the original 1,000 point series into a matrix of 6 columns by 500 rows. Each row represents a single exemplar and contains the data points for one time step in the first three columns, the predictor values, and the data points for the subsequent time step in the last three columns, the criterion values. This matrix is stored on disk as file *lorenz.trn*, and will be the training matrix for the network. In this way, the network will be presented repeatedly with three data points from the chaotic time series, one for each system variable as predictor values, and the corresponding following three data points as criterion values. For each of the three system variables the network will be required to learn to predict the value of the data point immediately following a given data point. A testing data set was constructed in the same way, starting from a new set of random initial values. This testing data is stored in file *lorenz.tst*.

Click the *Load Setup* button on a network dialog form to select the appropriate file from the following list:

Setup Files for Lorenz Flow Prediction Example

Network Function	Setup File
BackProp Network	lorenznn.set
Genetic Network	lorenzgn.set
Probabilistic Network	lorenzpn.set

Fractal Dimension Analysis

Introduction

The estimate of fractal dimension that will be discussed in this section is the correlation dimension. Correlation dimension (symbolized as d2) is an estimate of the lower bound on the number of variables involved in the dynamical behavior, or evolution over time, of a multivariate system. Correlation dimension can be computed by starting with time series measurements of each of the several variables of the multivariate system. Alternatively, it has been shown that under suitable conditions, an estimate of the correlation dimension can be computed from a univariate time series when that time series is an adequate sample taken from the multivariate process. The actual calculation of the correlation dimension has been reduced to a relatively straight-forward algorithm. For this reason, the correlation dimension has been applied as a measure of system complexity in a wide range of fields of endeavor.

Correlation dimension analysis could be used to analyze data recorded in a study in which several conditions are involved. A question that could be asked is, Is there a difference between the correlation dimensions of the data recorded in the different conditions?

The correlation dimension is one technique that can be used to estimate the complexity of a dynamical system. The term dynamical system can refer to any physical system that we choose to identify and demarcate. Examples of dynamical systems include a pendulum, an ensemble of neurons, and the solar system. Such a dynamical system is characterized by some pattern of behavior over time. This behavior may be relatively simple, as in a pendulum at rest or swinging freely. With such a system, what the system will do at future times can be predicted on the basis of information about the system's current behavior. The behavior of even simple dynamical systems can also be quite complex. An example is a pendulum with periodic forcing, a system whose future behavior may not be easily predictable. A dynamical system of this type, governed by clearly defined physical laws but whose future behavior is unpredictable from current behavior, is termed chaotic. Another example of a chaotic dynamical system is a billiard table with several balls in motion. The path of any one ball is determined by well-defined physical principles, and yet this path is not predictable for more than a short time following some given initial set of positions and velocities of the balls.

A consequence of the definition of a chaotic system is that the ongoing behavior of the system is sensitively dependent on its starting, or initial, conditions. Small changes to the starting positions or velocities of the billiard balls will after a short time cause the behavior of the system to diverge from the behavior of the undisturbed system. Such sensitivity to initial conditions is a defining characteristic of chaotic behavior. Figure 2.21 illustrates the phenomenon of extreme sensitivity to starting conditions. The logistic map was iterated to generate 2 sets of 40 data points. The starting point for the first set was 0.1, and for the second, 0.1001. Even with this relatively small difference in starting points the values generated by the logistic map differ after about data point 12.

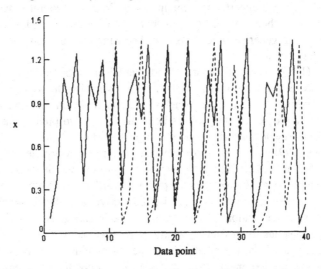

Figure 2.21. The graph shows sensitivity to initial conditions, illustrated by iterating the logistic map, $x(t + 1) = g \cdot x(t) \cdot (1 - x(t))$, with $g = 3$. The map was iterated twice, with starting values of 0.1 (solid line) and 0.1001 (dotted line) respectively. The behavior of this system is affected by the change in initial conditions after about twelve data points.

Chatterjee and Yilmaz (1992) offer an accessible review of chaos, its background, relation to statistics, and areas of application. They discuss the relationship between different estimators of attractor dimensionality, noting that these estimators, including the correlation dimension, are necessary but not sufficient conditions for an apparently chaotic dynamical process to be considered deterministic. The authors suggest that sufficiency conditions for labeling a chaotic-seeming process as deterministic have yet to be discovered. A slightly more technical treatment of the relationship between chaos, including the correlation dimension, and statistics is offered by Berliner (1992).

A measure that has been suggested as a necessary condition for a system to be considered chaotic is the Lyapunov exponent. Lyapunov exponents, described in more detail in a following section, are a quantitative index of sensitivity to changes in initial conditions. A positive Lyapunov exponent is considered to be evidence of deterministic chaos. Together, a fractional value of correlation dimension and a positive Lyapunov exponent would be compelling evidence for the presence of chaos.

Moon (1987) refers to Lyapunov exponents as being diagnostic of chaos. Jackson (1990) notes that there is wide agreement that a positive Lyapunov exponent must be present before a system can be considered to have a chaotic component. Farmer et al. (1983) provide a review of several different measures of attractor dimensionality, including the correlation dimension, and their relationship to Lyapunov exponents.

Chatterjee and Yilmaz (1992) note also that the discovery of chaos is evidence for the position that the question of whether an aperiodic process is deterministic or probabilistic may be undecidable. This question, they say, may join other such undecidable questions as the Heisenberg uncertainty principle and Godel's incompleteness theorem. Godel's theorem established that any system that is sufficiently complex, such as for instance, arithmetic, can generate statements that then cannot be proved correct or incorrect from within that system.

Ford (1987) points to Godel's fundamental theorem as the reason why chaotic dynamics are in principle unpredictable. Such dynamics, though resulting from possibly simple deterministic rules, nevertheless contain more information than can be encompassed by the logical system that is being used to try to predict the dynamical system's behavior. In essence, a part is trying to know the whole.

Hobbs (1991) offers an argument that attempts to reconcile what he considers to be the unnecessary debate between chaos and determinism. His thesis consists of two parts. First, he notes, chaotic processes are pervasive. All that is required for a system to exhibit chaotic behavior is that the system be dissipative, and that it be nonlinear. These two characteristics are seldom absent from natural systems. A swinging pendulum slowly coming to rest due to air friction would be an example of a dissipative system: Energy is being dissipated from the pendulum to the surrounding air. The second part of Hobbs' thesis is that the exponential sensitivity to initial conditions, a characteristic feature of chaotic systems, may allow these systems to amplify quantum fluctuations to a level where these fluctuations can have an influence on macroscopic phenomena.

Hobbs concludes that quantum-level indeterminism may become indeterminism at the macroscopic level. Determinism may in fact be no more than an illusion, rather than a reality, resulting from the particular scale at which phenomena are commonly observed. Determinism in the face of a chaotic universe would seem to require an infinite level of precision in knowing some initial state. Quantum uncertainty on the other hand does not allow for such infinite precision. The distant flapping of a butterfly's wings may indeed have an influence on local weather, but we could never know what the local weather would have been otherwise. Determinism may thus be less a principled, and more a human psychological requirement.

Lyapunov Exponents

A discussion of correlation dimension, raising the question of whether a dynamical system is behaving chaotically, brings in the concept of sensitivity to initial conditions. An index of how sensitive a dynamical system is to changes in starting conditions, or indeed to perturbations generally, is the Lyapunov exponent.

Consider the behavior of the logistic map illustrated in Figure 2.21. The logistic map was iterated twice, using slightly different initial values each time, 0.1 and 0.1001. After about 12 data points the 2 time series diverge. It turns out that for a chaotic system, as this one is, the rate at which the two time series diverge after some perturbation, or how quickly they become dissimilar, is a function of some positive power of the number of data points from the onset of the perturbation. This power is the Lyapunov exponent of this dynamical system.

To illustrate the idea of Lyapunov exponents, consider a dynamical system of a single variable. The logistic map is such a system. Suppose further that the difference between two time series generated by this hypothetical system was proportional to the square, or the second power, of the number of data points from the start of the series. Suppose that at time point 1 the two time series had the same value. At time point 2, then, they would differ by an amount proportional to 2^2, at time point 5 they would differ by an amount proportional to 5^2, and so on. Very roughly speaking, such a dynamical system would have a Lyapunov exponent of two. More precisely, Lyapunov exponents measure the rate of divergence between time series averaged over a number of points along the time series: A small difference at time-point 0, the initial value, produces some particular rate of divergence. A small difference, or perturbation, applied at time-point 5 might produce a somewhat different rate of divergence, and so on for further points on the time series. The Lyapunov exponent for this hypothetical dynamical system of one variable would be computed by averaging over these various rates of divergence. A dynamical system of n variables will have a set of n Lyapunov exponents. Perturbations made to each of the system variables will result in unique rates of average divergence. In other words, a dynamical system of variables x and y might be more sensitive to perturbations applied to variable x than to variable y. Taken together such a set of Lyapunov exponents, referred to as the Lyapunov spectrum of a dynamical system, describe the extent to which a dynamical system is sensitive to changes in initial conditions, and to perturbations generally.

If a dynamical system were not at all sensitive to such perturbations, the initial difference between evolutions of the system started from different initial values would not grow with time. If the initial difference was, say, 0.001, the difference after 10 or 100 data points would still be on the order of 0.001. This behavior could be expressed by saying that the difference was proportional to the zero-th power of the number of data points. The zero-th power of anything is just one. This amounts to saying that to get the difference at any time point, the initial difference is multiplied by one. Correspondingly, the Lyapunov exponent of this system would be zero. By the same reasoning, a negative Lyapunov exponent would mean that differences between evolutions of a system would actually shrink with time.

In sum, the larger the value of a positive Lyapunov exponent of a system, the more sensitive the system is to perturbations generally. And whatever the value of an exponent, if the exponent is positive, small perturbations will have an effect on the system that will increase rapidly as the system evolves. Such a system, with at least one positive Lyapunov exponent, can be considered to be chaotic.

Computational Considerations

Calculating the Correlation Dimension

One algorithm commonly used to compute correlation dimension for a time series was proposed by Grassberger and Procaccia (1983a, b). Calculation of correlation dimension begins by dividing up the original time series into sets of m-points, each of which is treated as a vector defining a point in an m-dimensional space. This m-dimensional space is often referred to as a phase space. The motivation for this step is a theoretical result proved by Takens (1981). Suppose that a single variable y is sampled repeatedly from the multivariate process defined by a vector x_t, to generate a time series y_t. The time series y_t is assumed to be an adequate sampling of a multivariate process. Typically, some components of x_t, that is some of the variables of the multivariate process, may be difficult or impossible to observe. The stipulation of adequate sampling requires that the variables of the multivariate process be sufficiently strongly coupled to the measured variable (Frank, Lookman, Nerenberg, Essex, Lemieux, and Blume, 1990). If this is so, Takens' (1981) result shows that multiple observations of the single variable y can be combined to create what can be treated as a single observation of the multivariate vector x_t. In more detail, the multiple observations of y are used to form vectors u_i as follows:

$$u_i = \{y_i, y_{i+t}, ..., y_{i+(m-1)t}\}$$

Here, t is the lag parameter, the spacing between adjacent observations that are combined to form the vectors u_i. These vectors define points in the m-dimensional phase space. Geometrically, the entire set of these points forms a pattern, termed an attractor, in the phase space. In the process of creating this attractor the time series is said to be embedded in the phase space.

Takens' (1981) result shows that under the right conditions, certain dynamical properties of the original multivariate process are preserved in this reconstruction of the original time series. A particular property that is preserved is the dimensionality of the original multivariate process. This dimensionality is reflected in the dimensionality of the attractor. One condition that must be met is that embedding phase space dimension m must be greater than or equal to 2D + 1, where D is the dimensionality of the attractor.

In the next step in the process of computing correlation dimension, the dimensionality of the phase space attractor is calculated. For this calculation, Grassberger and Procaccia (1983a, b) invoked the notion of the correlations between points on the attractor. These correlations are a function of the distance r around any one point on the attractor, and are measured by the correlation integral C(r). The value of the correlation integral, for some value of r, can be described as the proportion of the points on the attractor that are found within a distance r of a reference point on the attractor, averaged over all points on the attractor taken as reference points. The correlation integral C(r) is thus the average proportion of points on the attractor found within a volume of length r around all points on the attractor. The correlation integral is defined by the equation

$$C(r) = 1/n^2 \sum H(r - |y_i - y_j|) \hspace{2cm} 1$$

where the sum runs from i and j = 1 to n, n is the number of data points of the time series that are used for the calculation, and in the limit of large n; r is distance on the attractor; y_i and y_j are vectors defining points on the attractor. H is the Heaviside function defined by

$$H(x) = 0, x \leq 0$$

$$H(x) = 1, x > 0$$

The Heaviside function simply counts the number of pairs of points y_i and y_j that are separated by a distance less than r. The distance measure can be the usual Euclidean distance. The computation of Euclidean distance however is relatively time consuming. Alternative distance measures have been proposed that are faster to compute, and are equivalent for the purposes of calculating the correlation integral (Moon, 1987). The distance measure in equation 1 involves summing the absolute values of the differences between corresponding components of the 2 vectors in question. This measure is referred to as city-block distance: The distance between two points that one must travel following the constraints of streets organized around blocks. Using different measures of distance has the effect of changing the absolute range of the value of r in the graph of ln C(r) versus ln r. The slope of the graph is however unchanged.

Intuitively, at small values of r, that is in a small volume of phase space around any point on the attractor, there will be found on average a small fraction of the total points on the attractor. As the distance r increases this average fraction will increase. At sufficiently large values of r all points on the attractor will be found.

Grassberger and Procaccia (1983a, b) pointed out that the C(r) and r are related by the equation

$$C(r) = r^d \hspace{2cm} 2$$

where d is the correlation dimension. Thus, C(r) is a power function of r, in the limit of values of r that are small with respect to the size of the attractor. For instance, if d = 2, then C(r) increases as the square of r; the average proportion of points on the attractor that are found in the vicinity of any reference point on the attractor increases as the square of the distance around the reference point. This behavior is intuitively consistent with more familiar notions of dimension: For a two-dimensional surface, the size of the surface increases as the square of length along the surface.

In the case of an attractor however, the question of size becomes more complex than with most solid real-world objects because the attractor is composed of discontinuous points. Instead therefore, an analog of size is measured by computing the correlation integral C(r). The discontinuous nature of the attractor also leads to the notion of fractional values of dimensionality, in contrast with the more usual integer values of dimension assigned to physical objects that are geometrically continuous. Typically, attractors that are reconstructed from time series will have nonintegral values of (correlation) dimension. Such objects are termed strange attractors.

Solving equation 2 for d and taking the limit of small values of r, the dimensionality of an attractor can be expressed as

$$d = \ln C(r) / \ln r$$

The value of correlation dimension can thus be computed by finding the slope of the graph of ln C(r) versus ln r. Typically, this graph has an ogive, or S shape, so that a regression line is fitted to some intermediate portion of the graph. The slope of this regression line is then taken as the correlation dimension.

Factors Affecting Choice of Parameter Values

The computation of the correlation dimension involves the selection of a number of parameters that have been studied extensively in the last several years. See Dvorak and Siska (1986) for a discussion of the effect of these parameters on the value of correlation dimension.

Lag

A major source of ambiguity in the computation is the estimation of the value of the lag parameter used in forming the vectors that define the phase space attractor. The original Takens' formulation suggests that the value of lag used is not critical, given a time series that is both sufficiently long and noise free. In the practical case of a limited number of potentially noisy data points, the value of lag that is chosen for reconstructing the attractor in phase space is critical. With real-world data the value of correlation dimension is often dependent on the value of lag chosen for the reconstruction.

If we were to examine an attractor visually, with an unlimited amount of noise-free data, adjacent orbits on the attractor would be discriminable regardless of the value of lag. With a limited number of noisy data points the extent to which such orbits are distinguishable becomes a function of lag. At nonoptimal values of lag the attractor will be compressed along certain axes in the phase space. Noise, in the form of quantizing error and low-amplitude signals unrelated to the waveform of interest, will result in some orbits overlapping, and therefore being indistinguishable from adjacent orbits. At an optimal value of lag, the attractor will be most homogeneously distributed over all dimensions of the phase space. Orbits on the attractor will be maximally mutually discriminable. This observation forms the basis for the algorithm used to calculate lag in the *Correlation Dimension Analysis* function in *Simulnet*. At this optimum value of lag, the components of the resulting vectors will be maximally independent. In other words, a reasonable approximation to a set of independent basis vectors for the embedding phase space will have been found.

Numerous approaches have been proposed to try to deal with the issue of selecting an appropriate value of lag. Schaffer, Truty, and Fulmer (1988) have suggested that independent components for the embedding phase space are obtained when lag is chosen to be between 10% and 30% of the periodicity of interest in the time series.

Another often used approach to the problem of finding a value of lag that will produce orthogonal coordinates for the embedding phase space involves calculating the autocorrelation function for the time series, and then determining the first minimum of this function. The autocorrelation function is a measure of the linear dependence of either two variables, or a single variable with a lagged, or time-shifted, version of itself. A related approach (for example, Fraser and Swinney, 1986) involves the calculation of the mutual information function and its first minimum. Mutual information is a measure of the general dependence of two variables, or of one variable and a lagged version of itself. Fraser and Swinney (1986) argue in favor of mutual information over autocorrelation, claiming that mutual information will allow phase space coordinates to be found that are generally, and not just linearly, independent.

More recent work however has shown that neither autocorrelation nor mutual information is invariably successful in determining an optimal value of lag (Martinerie, Albano, Mees, and Rapp, 1992). Interestingly, Martinerie et al. (1992) found that the first minimum of the autocorrelation function was the most successful of several measures involving both the autocorrelation and the mutual information functions. The time series used for their study were the Rossler and Lorenz three-dimensional systems of differential equations, as well as a three-torus, a nonlinear oscillator driven at three incommensurate frequencies (see section 2.4). See also Leibert and Schuster (1989) for more detailed guidelines on the selection of the value of lag.

Embedding Dimension

A second potential source of ambiguity in the calculation of the correlation dimension is choosing the maximum value of embedding dimension m. Takens (1981) and Schaffer et al. (1988) suggest that m should be greater than $2n + 1$, where n is the dimensionality of the attractor. Without beforehand knowledge of n, m can be set initially at an arbitrarily high value (in practice, 10 to 15). The resulting estimate of correlation dimension can then be used to select a more appropriate value for m.

With a limited number of data points however, it has been observed that an overly high value for embedding dimension may result in an unstable estimate of correlation dimension: The scaling region, the straight-line portion of the graph of $\ln C(r)$ versus $\ln r$, becomes too short to permit a reliable estimate of its slope to be determined (Mayer-Kress and Layne, 1987).

Number of Data Points

Ruelle (1989) has argued for a minimum time-series length on the order of $a^{d/2}$, where a is proportional to the length of the scaling region; a is equal to the ratio of maximum to minimum ln r values over which the graph of ln C(r) versus ln r is approximately linear. The quantity d is an initial estimate of the correlation dimension of the time series. Ruelle suggests a value of a = 10. For attractors of high dimensionality and with a requirement for a long-scaling region, this criterion quickly leads to a need for an in-practice unrealizable number of data points.

Essex and Nerenberg (1991) have criticized these suggestions, claiming that Ruelle's bound applies to correlation dimension but not to the estimates made by computing the slope of the regression line to the intermediate segments of the ln C(r) versus ln r curve. Essex and Nerenberg suggest using a more modest scaling region requirement (for example, a = 2), and a more comprehensive calculation taking into account both the statistical and the geometric problems involved. Their calculation yields a value approximately 20 times smaller than Ruelle's requirement. Smith (1992) suggests that if accuracy requirements for the correlation dimension estimate are not too strict, then modest sample sizes are not unreasonable. For an RMS sampling error of 1, and an estimated dimension of 5, Smith (1992) suggests a sample size of 30. For an RMS sampling error of 0.1, sample size increases to about 5,000. Thus one effect of a small sample size is a larger RMS sampling error; in other words, a value of correlation dimension that is less stable with respect to sampling.

DeCoster and Mitchell (1991) investigated the behavior of the Grassberger-Procaccia algorithm as a function of the size of the data set, using data generated by systems of equations. They found that in some cases as few as 100 data points were sufficient to allow saturation of correlation dimension to be detected, indicating the presence of deterministic dynamics. In other cases even 5,000 data points were not sufficient to generate convergence, normally indicating that the time series in question represents a stochastic rather than a deterministic process. DeCoster and Mitchell concluded that the time series length requirement is not well defined, and that analysis should be attempted even when the number of available data points may be formally inadequate.

Using a small number of data points and not finding a stable value of correlation dimension, of course, leaves open the question of whether the data set does not represent deterministic chaos, or is simply too small. In one small sample study using economic data, Ramsey, Sayers, and Rothman (1990) were unable to find a stable value of correlation dimension. They could only conclude that there was no evidence for the presence of a chaotic attractor in their data.

Practical consideration often sets a limit on the maximum number of available data points; as, for example, when making EEG recordings. In the absence of pathology, the state of the brain can probably be assumed to remain stable for periods on the order of only a second or so. Time series recorded over intervals greater than several seconds, therefore, probably contain segments of differing statistical properties, and the time series would not be considered to be statistically stationary (Frank et al., 1990). With typical sampling rates of 100 to 300 samples per second, the number of available data points is on the order of several hundred. Sampling at greater rates will not alleviate this problem.

At increasingly high sampling rates, adjacent data points are increasingly correlated. Essentially, no new information is captured by sampling at greater than some optimal rate determined by the highest frequency of interest in the data. Such over-sampling can in fact have an effect on correlation dimension. Grassberger (1986) looked at the effect of interpolation on correlation dimension, finding that correlation dimension decreases with increasing amounts of interpolation.

The implication of these results for the analysis of small samples is the following: Although a small sample of, for example, several hundred points may be used to compute an estimate of the correlation dimension, this estimate might not be able to answer the question of whether or not motion on the attractor is chaotic.

Variable Transformations

When the phase space reconstruction leads to a highly nonuniform attractor, it may be difficult to calculate correlation dimension. The slope of the graph of ln C(r) versus ln r can, in this situation, have more than one straight-line region; and each of these regions may have a different value of slope.

In this situation, one approach has been to perform some sort of transformation on a dependent variable in the time series. Lefranc, Hennequin, and Glorieux (1992), for example, used the log transformation in a study of chaotic behavior in a modulated CO_2 laser. The result of the analysis of the transformed data was a more uniform attractor geometry, longer scaling regions, and saturation of the slope value. This approach assumes that relevant information is contained at all amplitudes of the time series, and hence, in all parts of the attractor. The log transformation makes this information more uniformly available to the correlation dimension calculation.

One technique commonly employed to increase the signal-to-noise ratio of measured data is averaging over multiple records. Assuming a noise component is uncorrelated across records, signal-to-noise ratio of the data increases as the square root of the number of records averaged. There is unfortunately a potential drawback to this procedure when the data is to be used to generate an estimate of correlation dimension. Badii, Broggi, Derighetti, Ravani, Ciliberto, Politi, and Rubio (1988) found that correlation dimension increases with averaging. An explanation for this effect might be that the averaging process somehow more closely couples variables from the different time series that enter into the average, thus increasing the dimensionality of the resulting averaged time series.

Attractor Geometry and Underlying Dynamics

Attractor geometry is by itself not a sufficient criterion to decide the question of whether or not the associated dynamical system is chaotic. Some attractors, generated by nonchaotic systems of equations, have been found which nevertheless have a fractal dimension value (Grebogi, Ott, Pelikan, and Yorke, 1984; Ding, Grebogi, and Ott, 1989; Romeiras, Bondeson, Ott, Antonsen, and Grebogi, 1987). Typically, the generating systems of equations consist of nonlinear oscillators driven at incommensurate frequencies. Attractors with fractional dimension values are referred to as strange attractors. A strange attractor does not then necessarily imply an underlying chaotic system.

Such results emphasize once more the point made earlier that inferences about the nature of the generating dynamical system cannot be made solely on the basis of the correlation dimension. In order to judge whether a dynamical system is chaotic, the extent to which the system is sensitive to changes in its initial conditions must be determined. This determination is usually made by calculating the Lyapunov exponents for the time series. Frank et al. (1990) for instance computed Lyapunov exponents for EEG recordings of epileptic seizure events. Their finding of positive exponents lends support to the notion that the underlying dynamics are deterministically chaotic. They suggest that the determination of chaos could not have been made without the calculation of Lyapunov exponents.

In a complementary study, Osborne and Provenzale (1989) found a class of stochastic systems with nevertheless finite values of correlation dimension. A purely stochastic system should not exhibit a finite value of correlation dimension. Dimensionality will in principle be equal to the number of data points. The particular class of systems used in this study were of the "colored noise" variety; the time series exhibited an inverse power-law spectrum. Such noise is common in a physical system, and is typically referred to as 1/f noise: The power spectrum decays with increasing frequency as the inverse of frequency. Intuitively, Osborne and Provenzale's (1989) finding results from the "coupling" of the individual degrees of freedom associated with the individual data points by the power-law function that defines the power spectrum. In the limit of a power-law function with a large exponent, high frequencies in the time series are effectively filtered out, and a sinusoidal function of correspondingly low dimensionality remains.

The implication of these results, for any data analysis using correlation dimension, is that it would be unsafe to draw inferences from the behavior of the correlation dimension about whether or not the underlying dynamics are or are not chaotic, and about the extent to which the time series represent deterministic dynamics or stochastic behavior. Pritchard and Duke (1992) emphasize the point that the Grassberger-Procaccia algorithm is most realistically useful, in a relative sense, in comparing systems for evidence of dissimilar complexity rather than in attempting to determine the absolute complexity of a single system. They advocate the term "dimensional complexity" rather than correlation dimension.

Alternative Methods to Calculating Correlation Dimension

A number of proposals have been put forward for calculating correlation dimension which attempt to deal with some of the practical difficulties that are involved, such as dealing with noisy data.

Broomhead and King (1986) have proposed an alternative to the Grassberger-Procaccia algorithm using singular value decomposition. This method attempts to reduce the effect of noise in the original time series on the calculation of correlation dimension, as well as circumventing the problem of choosing a value of lag. Their study demonstrated that this method provides an increase in the length of the scaling region in the plot of ln C(r) versus ln r, allowing more stable estimates of correlation dimension to be made.

The method proposed by Broomhead and King (1986) begins by using the method of lags to construct a sequence of n-dimensional vectors from the original time series. The set of these vectors constitutes a matrix X, from which an n by n covariance matrix X^TX may be formed (where T denotes the transpose). This covariance matrix may then be decomposed, giving a set of eigenvectors that form an orthonormal basis for an embedding space. By using only the m most significant eigenvectors (where m is, of course, less than n) to define the phase space, an attractor is created which presumably contains less noise than if the attractor were constructed using all n vectors to define the phase space. The dimensionality of this attractor is taken as being equal to the number of significant eigenvalues generated by the decomposition. According to Destexhe et al. (1988), however, this approach still suffers from the sensitivity to an adequate choice of the lag parameter. Gibson, Farmer, Casdagli, and Eubank (1992) have shown, however, that the number of significant eigenvalues is unrelated to the dimensionality of the attractor, but that the decomposition is nevertheless a useful procedure for improving the signal-to-noise ratio of the data.

Albano, Muench, and Schwartz (1988) combined the singular value decomposition method of Broomhead and King (1986) with the Grassberger-Procaccia algorithm. As before, singular value decomposition of the matrix of vectors formed using the method of lags is used to define an appropriate subspace using only the most significant eigenvectors. The resulting attractor is then subjected to analysis using the Grassberger-Procaccia algorithm. This method also gives a longer scaling region than does the Grassberger-Procaccia algorithm used alone, allowing more stable estimates to be made of the dimension value.

References

Albano, A. M., Muench, J., Schwartz, C., Mees, A. I., and Rapp, P. E. (1988). Singular-value decomposition and the Grassberger-Procaccia algorithm. *Physics Review A, 38, 3017-3026.*

Badii, R., Broggi, G., Derighetti, B., Ravani, M., Ciliberto, S., Politi, A., and Rubio, M. A. (1988). Dimension increase in filtered chaotic signals. *Physical Review Letters, 60 (11), 979-982.*

Berliner, L. M. (1992). Statistics, probability and chaos. *Statistical Science, 7 (1), 69-90.*

Broomhead, D. S. and King, G. P. (1986). Extracting qualitative dynamics from experimental data. *Physica 20D, 217-236.*

Casdagli, M., Eubank, S., Farmer, J. D. (1991). State-space reconstruction in the presence of noise. *Physica 51D, 352-359.*

Chatterjee, S. and Yilmaz, M. R. (1992). Use of estimated fractal dimension in model identification for time series. *Journal of Statistical Computation and Simulation, 41 (3/4), 129-141.*

Chatterjee, S. and Yilmaz, M. R. (1992). Chaos, fractals and statistics. *Statistical Science, 7 (1), 49-68.*

DeCoster, G. P. and Mitchell, D. W. (1991). The efficacy of the correlation dimension technique in detecting determinism in small samples. *Journal of Statistical Computation and Simulation, 39 (4), 221-229.*

Destexhe, A., Sepulchre, J. A., and Babloyantz, A. (1988). A comparative study of the experimental quantification of deterministic chaos. *Physics Letters A, 132, 101–106.*

Ding, M., Grebogi, C., and Ott, E. (1989). Dimensions of strange nonchaotic attractors. *Physics Letters A, 137 (4/5), 167-172.*

Dvorak, I. and Siska, J. (1986). On some problems encountered in the estimation of the correlation dimension of the EEG. *Physics Letters A, 118 (2), 63-66.*

Eckmann, J. P. and Ruelle, D. (1985). Ergodic theory of chaos and strange attractors. *Review of Modern Physics, 57, 617-656.*

Essex, C. and Nerenberg, M. A. H. (1990). Correlation dimension and systematic geometric effects. *Physics Review A, 42, 7065-7074.*

Essex, C. and Nerenberg, M. A. H. (1991). Comments on 'Deterministic chaos: the science and the fiction' by D. Ruelle. *Proceedings of the Royal Society of London A, 435, 287-292.*

Farmer, J. D., Ott, E., and Yorke, J. A. (1983). The dimension of chaotic attractors. *Physica 7D, 153-180.*

Ford, J. (1987). Directions in classical chaos. In Hao Bai-lin (Ed.), *Directions in Chaos.* Singapore: World Scientific.

Frank, G. W., Lookman, T., Nerenberg, M. A. H., Essex, C., Lemieux, J., and Blume, W. (1990). Chaotic time-series analyses of epileptic seizures. *Physica 46D, 427-438.*

Fraser, A. M. and Swinney, H. L. (1986). Independent coordinates for strange attractors from mutual information. *Physical Review A, 33 (2), 1134.*

Gibson, J. F., Farmer, J. D., Casdagli, M., and Eubank, S. (1992). An analytic approach to practical state-space reconstruction. *Physica 57D, 1-30.*

Grassberger, P. and Procaccia, I. (1983a). Characterization of strange attractors. *Physical Review Letters, 50 (5), 346-349.*

Grassberger, P. and Procaccia, I. (1983b). Measuring the strangeness of strange attractors. *Physica 9D, 189-208.*

Grassberger, P. (1986). Do climatic attractors exist? *Nature, 323, 609-612.*

Grebogi, C., Ott, E., Pelikan, S., and Yorke, J. A. (1984). Strange attractors that are not chaotic. *Physica 13D, 261-268.*

Hobbs, J. (1991). Chaos And Indeterminism. *Canadian Journal of Philosophy, 21 (2), 141–164.*

Jackson, E. A. (1990). *Perspectives of Nonlinear Dynamics, Vol. 2.* Cambridge: Cambridge University Press.

Lefranc, M., Hennequin, D., and Glorieux, P. (1992). Improved correlation dimension estimates through changes of variable. *Physics Letters A, 163 (4), 269-274.*

Leibert, W. and Schuster, H. G. (1989). Proper choice of the time-delay for the analysis of chaotic time series. *Physics Letters A, 142, 107-112.*

Martinerie, J. M., Albano, A. M., Mees, A. I., and Rapp, P. E. (1992). Mutual information, strange attractors, and the optimal estimation of dimension. *Physical Review A, 45 (10), 7058-7064.*

Moon, F. C. (1987). *Chaotic Vibrations: An introduction for applied scientists and engineers.* New York: John Wiley and Sons Ltd.

Osborne, A. R. and Provenzale, A. (1989). Finite correlation dimension for stochastic systems with power-law spectra. *Physica D, 35 (3), 357-381.*

Pritchard, W. S. and Duke, D. W. (1992). Dimensional analysis of no-task human EEG using the Grassberger-Procaccia method. *Psychophysiology, 29 (2), 182-192.*

Ramsey, J. B., Sayers, C. L., and Rothman, P. (1990). The statistical properties of dimension calculations using small data sets: some economic applications. *International Economic Review, 31 (4), 991–1020.*

Romeiras, F. J., Bondeson, A., Ott, E., Antonsen, T. M., and Grebogi, C. (1987). Quasiperiodically forced dynamical systems with strange nonchaotic attractors. *Physica 26D, 277-294.*

Smith, R. L. (1992). Comment: Relation between statistics and chaos. *Statistical Science, 7 (1), 109-113.*

Takens, F. (1980). Detecting strange attractors in turbulence. Dynamical Systems and Turbulence. *Lecture Notes in Mathematics, 898, 366-381.*

Simulnet Exercise: Computing Correlation Dimension

In this exercis correlation dimension will be computed for two matrices, each containing EEG data. The question to be answered is: what are the relative complexities of the dynamical systems whose behavior was sampled to create the data in the two matrices?

Given an unlimited amount of noise-free data, correlation dimension can estimate the number of variables (or alternatively, the number of degrees of freedom) that were involved in generating the data. Suppose that some dynamical system has been sampled to obtain the data. By computing the correlation dimension for this data, an estimate can be obtained of the complexity of the dynamical system in terms of the number of variables that are at work in the system. In the usual practical case where only a relatively limited amount of noisy data is available, the value of correlation dimension can not be interpreted as an estimate of the absolute number of variables in the generating system. Rather, correlation dimension can then only be used as an estimate of the relative degree of complexity of the underlying dynamical system whose behavior was sampled. For this reason, the results of the analysis in this exercise will be referred to as a dimensional complexity index, rather than as the correlation dimension.

This exercise will use the matrices in files *eeg01.dat* and *eeg10.dat*. Both of these files contain EEG data organized as 16 channels of 128 time-points. The data in these files was recorded in each of two conditions of an experiment in visual perception. File *eeg01.dat* was recorded in condition 1, while the subject was visually examining an image containing a camouflaged target object. File *eeg10.dat* was recorded in condition 2, while the subject was looking passively at a blank screen. As an initial hypothesis, we might assume that since condition 1 involves the relatively more complex neural processing (evaluating the image to discover a camouflaged target), file *eeg01.dat* should show the relatively higher value of correlation dimension.

Procedure

1. Close all forms on the desktop. Open file *eeg01.dat* and minimize the matrix form. This matrix will be referred to as the data matrix.

2. From the *Analyze* menu, select the *Correlation dimension* option. On the *Correlation dimension* dialog form, make the following changes:

   - Select the *Mode: Matrix* option.
   - In the *Distance: min* field, enter a value of 2.
   - In the *Distance: max* field, enter a value of 5.
   - In the *Distance: step* field, enter a value of .25.
   - In the *Regression: No. of points* field, enter a value of 3.
   - In the *Regression: min. coef.* field, enter a value of 2.

3. Click the *Compute* button. After a short interval, a matrix containing the results of the analysis will appear on the desktop. Next a graph form will appear containing a plot of the correlation integral as a function of distance. The plot will contain two labels; d2 and SE. The d2 label is the value of correlation dimension for the data matrix computed as the slope of the plot of the correlation integral at the tangent line shown on the graph in a contrasting color. The SE label is the standard error of regression; a measure of how well this tangent fits the data points on the correlation integral plot. This graph should resemble Figure 2.22 (the figure has been edited slightly for clarity).

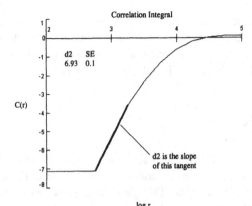

Figure 2.22. The graph shows the value of the correlation integral C(r), plotted as a function of the log of distance r. The flattest part of the curve, consistent with user-specified parameters, has been located, and its slope computed. This slope is d2, the estimate of the correlation dimension. The value of d2 for data matrix *eeg01.dat* is 6.93, with a standard error of regression (SE, a measure of the goodness of fit of the regression line to the curve) of 0.1.

4. Remove the data matrix and the results matrix *eeg1_1.dat* from the desktop. The desktop should now only contain the *Correlation dimension* dialog form and the graph form.

5. Open file *eeg10.dat*, and minimize the matrix form.

6. Click the *Compute* button. After a short interval a matrix containing the results of the analysis will appear on the desktop. This matrix will be called *eeg10_1.dat*. Next the graph form will be redrawn and a new value of d2 and SE will appear on the graph. This graph should resemble Figure 2.23 (the figure has been edited slightly for clarity).

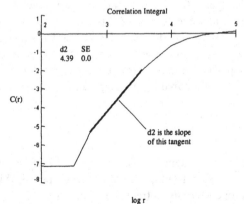

Figure 2.23. The graph shows the value of the correlation integral C(r) plotted as a function of the log of distance r. The flattest part of the curve, consistent with user-specified parameters, has been located, and its slope computed. This slope is d2; the estimate of the correlation dimension. The value of d2 for data matrix *eeg10.dat* is 4.39.

The estimate of correlation dimension for matrix *eeg01.dat* is 6.93, while for matrix *eeg10.dat* this estimate is 4.93. These results are consistent with our initial hypothesis that the complexity of the dynamical system sampled for the data in matrix *eeg01.dat* should be relatively higher than that for matrix *eeg10.dat*.

Fourier Analysis

Fourier analysis is a mathematical technique that allows a set of data points, representing some waveform, to be decomposed into the component frequencies that together add up to produce that waveform. This procedure produces a transformation of the original waveform, referred to as the Fourier transform of the waveform. Fourier transforms are generally computed using a computationally efficient algorithm referred to as the Fast Fourier Transform, or FFT. Using the Fourier transform, the spectrum of the waveform can be computed. The term spectrum refers to the 'spectrum' of frequencies that the Fourier analysis reveals as being present in the original waveform. As a visual analogy, a prism effectively performs a Fourier analysis on a beam of white light, decomposing the white light into its component frequencies. Visually, each of the component frequencies corresponds to a particular color. By examining the beam of light emerging from the prism, we can see what frequencies, or colors, were present in the original beam.

Simulnet Exercise: Computing a Fourier Spectrum

This exercise will use an already prepared matrix that contains data with known characteristics: noisy sine waves. The matrix consists of 8 columns, each of which has 512 rows. The data in each of these columns consist of 16 cycles of a sine wave to which pseudo-random noise has been added. Assuming for the purpose of this exercise that these 16 cycles span a time of 1 second, the sine waves have a frequency of 16 Hz. In the following procedure, one column of this matrix will be Fourier analyzed. The result will be the spectrum of the data in this column. The term spectrum is more precisely referred to as the power spectrum. Power in this context can be considered to be equivalent to variance. The power spectrum shows the relative amounts of power, or variance, that are contained in each of the frequency components included in the spectrum.

The Research Question

What are the relative magnitudes of the frequency components into which the data can be decomposed? An associated question that could be asked is: what are the corresponding phase angles for each of these frequency components?

Procedure

1. Close all forms on the desktop. Open file *sinnois2.dat* and minimize the matrix form.

2. From the *Analyze* menu, select the *Fourier Transform* option. On the *Fourier Transform* dialog form, enter 1 in the *Data matrix: End column* field.

3. Click the *Compute* button. A matrix form containing the spectrum will appear on the desktop. Close the *FFT* dialog form. There should now be two matrix forms on the desktop, the original matrix *sinnois2.dat* and its power spectrum *sinnoi_1.dat*. Using the following graphing procedures, create the following two graphs:

(1) Graph of column 1 of matrix *sinnois2.dat*:

   • Click the matrix button corresponding to this matrix (it should be button 1).
   • From the *Graph* menu, select the *XY Graph* option.
   • On the *XY Graph* dialog form, enter 1 in the *Y axis: End column* field.
   • Click the *New Graph* button.

   A new graph will be created that should resemble Figure 2.24. Examine this graph and note that there are no easily evident regularities; the graph shows 16 cycles of sine wave with added pseudo-random noise.

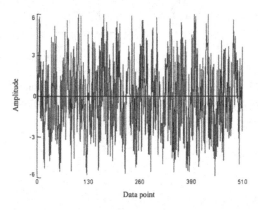

Figure 2.24. The graph shows column 1 from matrix *sinnois2.dat*, consisting of a 16 Hz sine wave, with added pseudo-random noise.

(2) Graph of the power spectrum, *sinnoi_1.dat*:

   • Click the matrix button corresponding to this matrix (this should be button 2).
   • On the *XY Graph* dialog form, enter 1 in the *Y axis: End column* field.
   • De-select the *X axis: Use row no.* option, and enter 2 in the *X axis: Use column* field (the reason for these settings is that the spectrum matrix uses column 1 to hold the values of power and column 2 to hold the values of frequency).
   • Click the *Redraw* button.

The new graph should resemble Figure 2.25. Inspecting the graph of the power spectrum we can easily see a relatively large power component at 16 Hz. This component rises clearly above a noise floor that is evenly distributed in power over the entire range of frequencies.

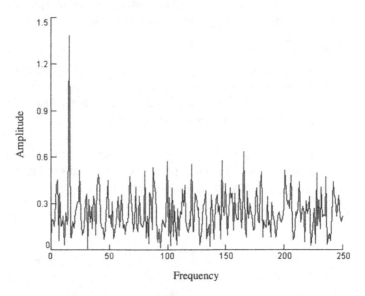

Figure 2.25. The graph shows the power spectrum of column 1 of matrix *si-nois2.dat*. The pseudo-random noise is evenly distributed over all frequencies, above which the 16 Hz sine wave component can easily be seen.

Eigenvalue Analysis

Introduction

The following discussion of eigenvalues and eigenvectors will be based on geometrical descriptions rather than the more traditional algebraic development of this topic. This approach has been adopted for two reasons. The first is that this discussion is aimed at data analysts who want to use this procedure to analyze time-series data to estimate the complexity of the sampled system. For such users geometric, rather than algebraic, interpretation of eigenvalues and eigenvectors is probably more appropriate. The second reason for using a geometric approach is to make the topic more accessible to data analysts with the relatively modest mathematical background that has been assumed. There are many texts available that develop this topic algebraically. See for example Fraleigh and Beauregard (1990).

Estimating System Complexity

Eigenvalue analysis is a technique for estimating the relative amount of variance in a data matrix, along each of n dimensions. Here n is the number of variables whose activity was sampled in creating the matrix. Generally this is assumed to be equal to the number of columns in the data matrix, each column containing the data for one variable. In any event, each of the variables in the data matrix can be thought of as corresponding to one of the dimensions of the data.

Picture the data as a group of points in an n-dimensional space. This group of points will have an extent along each of the n dimensions. The extent along any one dimension is proportional to the variance within that data matrix with respect to the variable corresponding to that dimension. In turn, the variance along any one dimension is measured by a quantity referred to as an eigenvalue of the data matrix. Eigenvalues thus simply measure the amount of variance within the data along a particular dimension. For a matrix of n columns, or n dimensions, there will always be n eigenvalues, one for each dimension. If the n variables whose behavior was sampled are mutually independent, then the corresponding n eigenvalues will all be distinct.

There is not, however, a direct correspondence between any one eigenvalue and any particular matrix column. In the original matrix, each column, containing the data for one variable, has associated with it a variance. Through the procedure used to obtain the eigenvalues, the eigenvalue analysis, this variance is 'repackaged': The original variables are used to create a set of new variables which can be thought of as new dimensions, each of which is a linear combination of the original variables or dimensions.

The way in which these new variables are created from the original variables, the weights used to combine the original variables, is specified by the (rescaled) elements of what are termed eigenvectors. Eigenvectors can be thought of as the 'directions' in the original data that lie along the major axes of the data. Imagining that the data are plotted as a group of points in a three-dimensional space, then in general, the shape of this group can be approximately described as an ellipsoid, an ellipse in three dimensions. This ellipsoid will have a major axis, the direction along which the ellipsoid extends most. This direction is defined by one of the eigenvectors of the data matrix. For every eigenvector there exists a corresponding eigenvalue. The eigenvalue specifies how far the group of points extends along the particular direction defined by its associated eigenvector. This ellipsoid will have one longest axis, corresponding to the largest eigenvalue and its associated eigenvector. The second-longest axis of the ellipsoid, normal to the first axis, corresponds to the second-largest eigenvalue along with its associated eigenvector. In this example of a thee-dimensional ellipsoid there will be one final axis, the third-longest axis, normal to both of the first two axes. This third axis corresponds to the third-largest eigenvalue of the data matrix, again with an associated eigenvector.

The significance of these new variables or dimensions defined by the eigenvectors is that they are mutually, linearly independent. The original data dimensions may or may not have been mutually, linearly independent. Since they are mutually independent, these new dimensions can serve as a coordinate system upon which the data can be plotted. Such a plot would show how many dimensions the data does actually occupy. The plot might reveal, for instance, that the data in a matrix of three columns, purportedly corresponding to three dimensions or variables, actually occupies a space of only two dimensions. This would imply that the variables corresponding to the three columns are not mutually independent. Rather, the data in some columns could be shown to be a linear combination of the data in other columns.

To visualize this situation, consider again a matrix of three columns (the number of rows is irrelevant). This matrix is assumed to have been created by sampling the activity of three measurement variables, each column representing a single variable. The data in the matrix can be pictured as a group of points in a three-dimensional space. Each row of the matrix gives us a set of three numbers that define a single point in this space. The data in the three columns may or may not be mutually independent. If the columns are not all independent of each other, then the axes that define the thee dimensions in the matrix will not lie at right angles to each other. Only if the matrix columns are mutually independent will the associated dimensions lie at right angles. In this case the dimensions are said to be orthogonal. The data matrix is then also said to be orthogonal. The cases of three, two, and one independent columns of data will now be considered in turn.

Let us assume first that the three variables in the matrix are mutually independent. The behavior of any one variable does not affect, and is not affected by, any other variable. Let us assume also that the variance of the three columns is equal. If we were to graph these variables on a three-dimensional plot, we would find that the points define a shape that extends approximately equally in all directions. The eigenvalues of this matrix provide a description of this shape. The size of the eigenvalues is an index of the extent of this shape along the three dimensions of the space. In this example, with the three mutually independent columns having equal variance, we should expect that the eigenvalues will be of approximately equal size. The graph shown in Figure 2.26, and the corresponding eigenvalues listed in column 1 of Table 2.5, illustrate this situation.

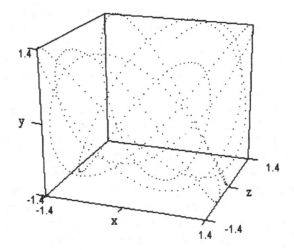

Figure 2.26. The volume of space occupied by the data points extends approximately evenly in all directions. These data points thus occupy a three-dimensional space.

Table 2.5 Eigenvalues versus Number of Independent Columns

3	2	1
1.196	2.041	3.011
0.803	1.003	0
1.032	0	0

Let us assume next that the three-column matrix contains data representing the activity of only two variables. Column 1 contains data that represents the behavior of the first variable, but the data in columns 2 and 3 both represent the activity of the second variable. Essentially, the data in these columns is highly, actually perfectly, correlated. The graph of this matrix is shown in Figure 2.27, and the corresponding eigenvalues are shown in column 2 of Table 2.5. The graph shows that the data points occupy a horseshoe-shaped region. This region lies within a plane, a two-dimensional subspace of the three-dimensional volume. We should expect that there will then be only two non-zero eigenvalues associated with this matrix. Column 2 of Table 2.5 shows that this is so. Note that eigenvalue analysis always gives one eigenvalue for each column in the original data matrix, but that some of the eigenvalues may be very small or zero as in this example.

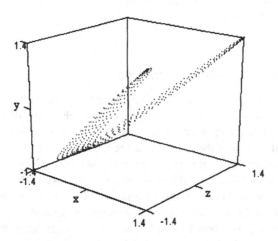

Figure 2.27. The volume of space occupied by the data points lies on a plane, a two-dimensional subspace of the three-dimensional volume.

Let us assume finally that the three-column matrix contains data representing the activity of only one variable. Columns 1, 2, and 3 all represent the behavior of this single variable. The data in these three columns is thus, of course, highly correlated. Figure 2.28 shows a graph of this matrix. The graph shows that the data points lie on a line, a one-dimensional subspace of the three-dimensional volume. We should expect then that the matrix has associated with it only one non-zero eigenvalue. Column 3 of Table 2.5 shows this to be the case.

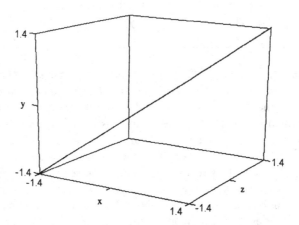

Figure 2.28. The volume of space occupied by the data points lies on a line, a one-dimensional subspace of the three-dimensional volume.

Creating Independent Variables

The preceding discussion can be summarized as follows: The data matrix consists of columns, each of which contains data that was sampled from one variable of the system that was investigated. Depending on the choice of factors, such as how the variables were selected and then how they were measured, these measured variables may or may not be mutually independent. It might be the case that some of these variables are actually linear combinations of other variables. In that case, there would be fewer real variables then the number of columns in the data matrix would suggest. Computing eigenvectors and eigenvalues for this matrix provides a prescription for how the variables in the data matrix may be linearly combined to yield a set of new variables that are mutually independent. This number might, in some cases, be less then the number of measured variables.

Eigenvectors describe how the linear combinations of the matrix variables are formed so as to create these new and independent variables. In particular, the eigenvector corresponding to the largest eigenvalue defines how the data matrix variables are to be combined so as to form a new variable that will account for the largest possible portion of the variance in the data. The size of this largest eigenvalue is an index of the magnitude of this variance. This process of defining the new, mutually independent variables is repeated for the eigenvector corresponding to the next largest eigenvalue, and so on. In some situations, it becomes important to know just how to recombine a set of measured variables to create a new set of synthetic but mutually independent variables.

The Research Question

While a matrix represents the behavior of as many variables as there are columns in the matrix, not all of these variables may be mutually independent. Some may be linear combinations of the other variables. Given this possibility, how many independent variables, or in alternate terms, orthogonal dimensions, are required to describe the data?

Simulnet Exercise: Computing Eigenvalues

In this exercise the eigenvalues for a matrix of three columns of data will be computed. The columns contain sine waves with frequencies that have been chosen to be mutually incommensurate. In other words, no pair of frequencies can be expressed as the ratio of integers. What this means is that the data in the three columns is mutually independent: None of the columns can be expressed as a linear combination of the data in the other columns. The data was generated using the equations

$$f(x) = a \cdot \sin(2\pi f_1 x)$$

$$f(y) = a \cdot \sin(2\pi f_2 y)$$

$$f(z) = a \cdot \sin(2\pi f_3 z)$$

where $a = 1.4$, and $f_1 = 3.414$, $f_2 = 5.567$, and $f_3 = 7.133$. Plotting the data on an xyz graph (Figure 2.29), it appears that the data points lie on a three-dimensional surface. The amplitudes of the three sine waves, determined by the value of parameter a, are equal. As a consequence, we would expect that this matrix will have three eigenvalues of approximately equal size.

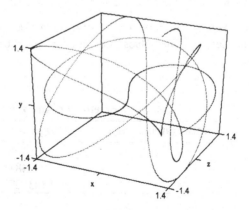

Figure 2.29. The data points generated by three sine waves with mutually incommensurate frequencies (file *evexamp.dat*) occupy a three-dimensional space

Procedure

1. Close all forms on the desktop. Open file *evexamp.dat* and minimize the matrix form. This matrix contains 3 columns of 500 rows. Each column contains a sine wave. The frequencies of the three sine waves are mutually incommensurate. This matrix is shown in Figure 2.30.

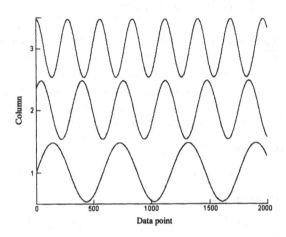

Figure 2.30. The graph shows the matrix in file *evexamp.dat*. The y axis labels indicate the column number of the matrix. The three columns each contain a sine wave of a different frequency.

2. From the *Analyze* menu, select the *Eigenvectors/values* option.

3. Click the *Compute* button. A matrix form containing the eigenvalues matrix will appear on the desktop. The values in the matrix should be close to those in the following table.

Results of Eigenvalue Analysis

Data matrix column	Eigenvalue
1	1.176196
2	0.9994814
3	0.8432291

As expected the three eigenvalues are approximately of equal size. This result indicates that the original data matrix contained data sampled from three independent variables with approximately equal variance for each variable.

References

Fraleigh, J. B. and Beauregard, R. A. (1990). *Linear Algebra*. Reading, Mass.: Addison-Wesley.

Coherence and Phase Analysis

Introduction

Coherence is one measure of the degree of association between two variables. For the purposes of this discussion, we will assume that the behavior of these variables is time-dependent (although in general, this need not be the case), and that the values of each of the variables have been recorded over some interval of time. Now, in order to describe coherence, it might be easiest to use an analogy with correlation. The coefficient of correlation is a measure of the degree of linear association between two variables. The degree of correlation is estimated by computing the correlation coefficient. The correlation coefficient can be either positive or negative. If the respective changes in the behavior of the two variables occur in the same relative direction, correlation is positive; if in opposite directions, correlation is negative. The square of the correlation coefficient is also a measure of linear association. Because of the squaring operation, squared correlation is always positive and thus independent of the relative direction of the changes in the data samples. Squared correlation, also referred to as the coefficient of determination, indexes only the strength of a linear association: how much of the behavior of one variable can be predicted from the behavior of the second variable.

Returning now to coherence: Coherence can be considered as the frequency domain analog of squared correlation. Squared correlation indicates the linear association between, in this discussion, time-related changes in the behavior of a pair of variables. Coherence, on the other hand, indicates the extent of linear association between frequency-related changes in the behavior of the two variables. This difference between the two measures can be expressed by stating that correlation is defined in the time domain, while coherence is defined in the frequency domain.

The second measure generated by coherence analysis is phase. Graphically, phase is a measure of the degree to which a pair of waveforms 'lines up'. That is, phase is an index of the extent to which the features of the two simple waveform, such as a sine or cosine, coincide in time. If the waveforms are sine waves, then phase indicates the extent to which the peaks and troughs of the sine waves coincide. Phase is measured in some unit of angular measure, such as degrees of radians.

More technically, coherence is the correlation between the complex Fourier power spectra computed for each of two time series. This correlation between the two power spectra is referred to as the cross spectral density (CSD). As a practical matter, coherence is computed by averaging the value of CSD over some range of frequencies. Coherence can be considered to be an estimate of the amount of shared power, or variance, within that frequency range, between the two time series. Thus, we consider the power, or variance, within some specified frequency band in each of the two time series. We might, for example, consider the variance in the 2 to 6-Hz frequency band. Coherence represents the proportion of this variance in one of the time series that can be accounted for by a linear function of the variance in the other time series (Otnes and Enochson, 1972, 1978). In these terms, the analogy with squared correlation, or the proportion of the variance of one variable that can be accounted for by a linear function of a second variable, may be a little clearer.

Practical Issues

A consideration when using coherence analysis is the volume of data that is available. The coherence computation is relatively demanding in terms of the number of data points that are needed to produce a reliable estimate. Let us assume that coherence is being computed between a pair of time series, each of length n. The maximum number of frequency components into which these time series can be decomposed is n / 2. Thus, if the time series each consist of 256 data points, they can be Fourier analyzed into 128 (256/2) frequency components. Coherence is computed by calculating the CSD between these two time series. This calculation involves averaging the CSD over some number of the frequency components. The coherence computation might, for instance, involve averaging over eight frequency components. Estimates of CSD are distributed approximately as chi-squared variables (Otnes and Enochson, 1972), with a standard error of estimate SE given by $1 / \sqrt{n}$, where n is the number of individual frequency values averaged over in computing the CSD. For instance, if the average was taken over eight frequency components, SE becomes

$$1 / \sqrt{16} = 0.25.$$

Standard error SE can be translated into a confidence interval for the coherence estimate. A confidence interval is a range of values of some statistic, in this case coherence, which can be assumed to include the true value of the statistic with some degree of confidence. Confidence intervals are computed using the equation

$$CI = value \pm z_{\alpha/2} \, SE$$

where *value* is the value of the statistic in question; $z_{\alpha/2}$ is the z statistic, and has a value equal to 1.96 for a 95% confidence interval, and to 2.58 for a 99% confidence interval; and SE is the standard error of estimate of the statistic, 0.25 in the present example. Suppose that a value of coherence of 0.75 has been computed. The 95% confidence interval then becomes

$$CI = 0.75 \pm (1.96)(0.25) = 0.75 \pm .49 = [0.26, 1.24].$$

In practical terms, the true value of coherence lies within the range of 0.26 to 1.24 with a confidence of 95%. This is clearly a wide range of coherence value. In order to generate a coherence value with a narrower confidence interval, the value of standard error of estimate SE must be decreased. In turn, this means computing the coherence value by averaging over more than eight frequency components. In the extreme, averaging over all 128 frequency components that are available, the standard error SE becomes $1 / \sqrt{128}$ = 0.088, resulting in a confidence interval of

$$CI = 0.75 \pm (1.96)(0.088) = 0.75 \pm .173 = [0.58, 0.92].$$

This is the narrowest confidence interval of the value of coherence that can be computed given the 128 frequency components obtained from the original 256 data points. A still narrower confidence interval would require more data points. We see from the equation defining confidence interval that the width of the interval is inversely proportional to the square root of the number of available frequency components. In other words, in order to decrease the width of the interval by a factor of two, to make it half as wide, the number of data points must be increased by a factor of four, and so on.

The overall message is that reliable estimates of coherence require a minimum of several hundred, and ideally several thousand, data points. In comparison, measures of association, such as correlation, require far fewer data points for a reliable estimate.

Computation

The coherence computation first involves computing complex Fourier spectra for two time series, $X(f)$ and $Y(f)$. Next, from these Fourier spectra, the following quantities are computed:

$$Gx(f) = (2 / n) \, |X(f)|^2, \text{ the power spectral density of } X(f)$$

$$Gy(f) = (2 / n) \, |Y(f)|^2, \text{ the power spectral density of } Y(f)$$

$$Gxy(f) = (2 / n) \, [X^*(f) \, Y(f)], \text{ the cross spectral density}$$

where $| \, . \, |$ denotes the absolute value and * denotes the complex conjugate. From the cross power density, the cospectra $Cxy(f)$, and quadspectra $Qxy(f)$ are computed using the relation

$$|Gxy(f)|^2 = Cxy(f)^2 + Qxy(f)^2$$

The cospectra and quadspectra represent, respectively, the real and imaginary components of the cross power density. Next, coherence γ and phase ϕ are computed for each frequency component. Coherence is computed by dividing the squared absolute value of the cross power density by the power spectral densities of the two time series, a normalizing operation. Phase is computed by calculating the inverse tangent of the ratio of the quadspectrum to the cospectrum.

$$\gamma(f) = |Gxy(f)|^2 / [Gx(f) \, Gy(f)]$$

$$\phi(f) = \arctan (Qxy / Cxy)$$

Finally, smoothed values of coherence and phase are computed by averaging over a range of n frequency components.

$$\text{Coherence} = (1 / n) \sum \gamma_i$$

$$\text{Phase(f)} = (1 / n) \sum \phi_i$$

This value of coherence represents the average cross-correlation between the power spectra of the two time series, normalized by dividing by the respective power spectral densities for the two individual time series. The values of phase are specified in degrees.

The Research Question

What is the coherence between the two sets of data, each representing the behavior of a single variable? That is, to what extent are the frequency components within the two data sets correlated?

Simulnet Exercise: Computing Coherence and Phase

In this exercise the coherence between each pair of columns in a matrix that contains 3 columns by 512 rows will be computed. Columns 1 and 2 contain 16 cycles of a sine wave with added pseudo-random noise, while column 3 contains only noise. This exercise will assume that the data in each of these columns represents an interval of 1 second. The data in columns 1 and 2 can therefore be referred to as sine waves with a frequency of 16 Hz.

Procedure

1. Close all forms on the desktop. Open file *sinnois3.dat* and minimize the matrix form. This matrix is shown in Figure 2.31.

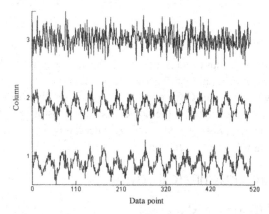

Figure 2.31. The graph shows the matrix in file *sinnois3.dat*. Columns 1 and 2 contain noisy sine waves, while column 3 contains only pseudo-random noise.

2. From the *Analyze* menu, select the *Matrix Functions* option. On the *Matrix Functions* dialog form, click on the *Operation* list-box to show the list of available options, and scroll through the list to find the *Coherence* function. In the *Data Matrix: End column* data entry field, enter a value of 2. In the *Parameters: Sampling rate* data entry field enter a value of 512. In the *Parameters: Averaging Interval* data entry field enter a value of 2. These settings will allow the coherence to be computed between data matrix columns 1 and 2.

3. Click the *Compute* button. A matrix form containing the results matrix will appear on the desktop. This matrix contains the following information; column 1 contains values of coherence, column 2 contains values of phase, and column 3 contains the frequency scale. The first few and the last few rows of the matrix are shown in the following table:

Results of Coherence Analysis: Columns 1 and 2

Coherence	Phase (degrees)	Frequency (Hz)
0.0368	-16.6	3.5
0.0671	89.1	11.5
0.976	-1.87	19.5
.	.	
0.0183	-79.2	236
0.392	-2.28	244
0.047	84	252

A convenient way to view the values of coherence is to plot them on an xy plot. Figure 2.32 shows a graph of column 1 on the y axis, and column 3 on the x axis.

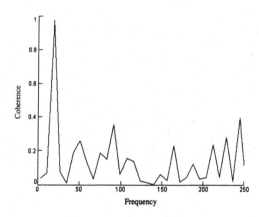

Figure 2.32. The graph shows the coherence for columns 1 and 2 of the matrix in file *sinnois3.dat*. Coherence peaks at approximately 16 Hz, the frequency of the noisy sine waves in columns 1 and 2 of the matrix.

Since columns 1 and 2 of the data matrix contains noisy sine waves with a frequency of 16 Hz, we should expect a high value of coherence at this frequency. That is, the 16-Hz frequency component in columns 1 and 2 should itself be highly correlated. Since coherence is the degree of correlation between frequency components, there should correspondingly be a high correlation at 16 Hz. As Figure 2.32 shows, there is a coherence peak at approximately this frequency. The reason that the peak of the coherence value does not fall exactly at 16 Hz is that the coherence computation involved forming an average over blocks of frequency components. Essentially, a measure of the correlation between all frequency components was computed individually, from which the value of coherence was computed by averaging this measure of correlation across groups of eight frequency components. The result is values of coherence for relatively coarse frequency steps.

4. Click the matrix button corresponding to the original data matrix, *sinnois3.dat*, to set the focus to this matrix. On the *Matrix Functions* dialog form, in the *Data Matrix: End column* data entry field, enter a value of 3 (it may already be 3). These settings will allow the coherence to be computed between data matrix columns 1 and 3.

5. Click the *Compute* button. A matrix form containing the results matrix will appear on the desktop. Plotting column 1 of the results matrix on the y axis, and column 3 of the results matrix on the x axis, should produce a graph resembling Figure 2.33.

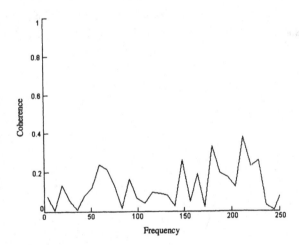

Figure 2.33. The graph shows the coherence for columns 1 and 3 of the matrix in file *sinnois3.dat*. Since column 1 contains a noisy sine wave while column 3 contains only noise, coherence values are therefore low at all frequencies.

Questions:

1. Why does the coherence corresponding to the 16 Hz frequency component actually appear in the results matrix at a frequency of 19.5 Hz?

2. What could account for the result that coherence at frequencies other than at 16 Hz is not equal to 0?

3. How does the value of phase for the 19.5 Hz frequency step compare with the phase for other frequency steps?

Answers:

1. Because of the averaging process that is used as part of the coherence computation, the results matrix contains coherence values for only relatively coarse frequency steps. The closest frequency in the results matrix to 16 Hz happens to be 19.5 Hz.

2. The data matrix contains sine waves together with added pseudo-random noise. The chance coherence between the frequency components of this noise is responsible for the observed non-zero values of coherence for frequencies other than that corresponding to the frequency of the sine waves.

3. The phase of the 19.5-Hz frequency step reflects, to a large extent, the phase of the 16-Hz sine waves in the data matrix. The phases of the other frequency steps reflect entirely the pseudo-random noise in the data matrix. Since the 16 Hz sine waves were designed to 'line up', that is the peaks and troughs of these sine waves were designed to coincide, the phase for the 19.5-Hz frequency step is expected to be small. Examining the table of coherence results we can see that the 19.5-Hz frequency step has the smallest value of phase, 1.87 degrees. The sign of the phase angle is disregarded since whether a phase angle is positive or negative simply indicates whether one signal leads or lags the second signal. How much it leads or lags is what is relevant at this point, and this is indicated by the absolute value of phase.

    The phase of the 19.5-Hz frequency step is smaller than the phase of any of the other frequency steps. These other steps correspond to noise components in the data. Since phase indicates the relative extent to which the frequency components in the 2 columns 'line up', that is the extent to which the peaks and troughs of these components coincide, we should expect that the phase of the 19.5-Hz frequency step should be relatively small. This frequency step corresponds to the 16-Hz sine waves, and these sine waves were designed to 'line up', that is to have a 0 phase.

References

Otnes, R. K. and Enochson, L. D. (1972). *Digital Time Series Analysis*. New York: John Wiley & Sons.

Otnes, R. K. and Enochson, L. D. (1978). *Applied Time Series Analysis*. New York: John Wiley & Sons.

Mutual Information Analysis

Introduction

Mutual information, based on the information theory developed by Claude Shannon (1948), is an index of the degree of association between two variables. More particularly, mutual information is a measure of how much information about one variable can be predicted by making a measurement of the second variable. In contrast with correlation that estimates the strength of the linear relationship between two variables, mutual information estimates the strength of the general relationship between them. By comparing the results of the mutual information and correlation analysis, it may be possible to estimate how well the relationship between two variables can be considered to be linear.

Computation

In order to compute mutual information, we define an event x as some interval of values of the data. Suppose that the total range of values in the data is -80 to +80. This range is divided into some number of intervals. Choosing 8 intervals, the first ranges from -80 to -70, and the last from +70 to +80. The data is then binned by assigning each data point to one of the intervals. Next, for each interval a probability $P(x)$ is computed; $P(x)$ is the probability that the data contains a value lying within interval x. Each of the probabilities is multiplied by the logarithm of that probability, $P(x) \log P(x)$. These products are then summed over the total number of intervals giving $\Sigma P(x) \log P(x)$. This sum is an estimate of the entropy of the system from which the data was sampled; the average amount of information derived from a single measurement made on the system. When the logarithm is taken to base 2, the units of entropy are bits.

In more detail, the information content $I(x)$ of an event x is related to its probability $P(x)$. The less probable the event, the higher its information content according to the equation

$$I(x) = -\log P(x)$$

The average information content for a set of events X is computed as

$$H(X) = \Sigma P(x)I(x)$$

This quantity, referred to as the Shannon entropy or just entropy, is the information content of an event x, weighted by its probability, and summed over all of the events in the set. Combining these two equations, entropy is computed as

$$H(X) = - \sum P(x) \log P(x)$$

Mutual information can now be defined in terms of entropy (Fraser and Swinney, 1986; Gray, 1990). Consider two sets of events X and Y. The value of entropy for each of these sets of events, H(X) and H(Y), can be computed as described. Next, the concept of entropy is extended to include the case of a pair of events, x and y, from these two systems. The joint entropy H(X, Y) is the amount of information available from this single pair of measurements of systems X and Y.

To compute joint entropy, pairs of events x and y from the two sets are first binned. Continuing the example above, each of the individual measurements from systems X and Y are placed into intervals of 10 units, beginning with the -80 to -70 interval and ending with the 70 to 80 interval. These separate intervals are combined to form a discrete joint frequency distribution of n by n bins. The first bin, for instance, contains those measurements for which events x and y both fall within the range of -80 to -70 , and so on. For each of these bins a joint probability P(x, y) is computed to create a discrete joint probability distribution for the two sets of events. Finally, the sum of the products of these probabilities and their logarithms is accumulated. The joint entropy of systems X and Y is computed as

$$H(X, Y) = - \sum P(x, y) \log P(x, y)$$

Mutual Information is then defined in terms of the individual entropies of systems X and Y, and their joint entropy as the sum of the individual entropies minus their joint entropy:

$$I(X, Y) = H(X) + H(Y) - H(X, Y)$$

In the present analysis, all logarithms are taken to base 2. The resulting values of mutual information then represent the number of bits of information that can be predicted about one time series from a measurement made on a second time series.

The Research Question

What is the value of mutual information between a pair of variables? That is, to what extent can the behavior of the one variable be predicted from measurements made on the other variable?

Simulnet Exercise: Computing Mutual Information

In this exercise the mutual information between each pair of columns in a matrix that contains 3 columns by 512 rows will be computed. Columns 1 and 2 contain 16 cycles of a sine wave with added pseudo-random noise while column 3 contains only noise. This exercise will assume that the data in each of these columns represents an interval of 1 second. The data in the columns 1 and 2 can then be considered to be sine waves with a frequency of 16 Hz.

Procedure

1. Close all forms on the desktop. Open file *sinnois3.dat* and minimize the matrix form.

2. From the *Analyze* menu, select the *Matrix Functions* option. On the *Matrix Functions* dialog form, click on the *Operation* list-box to show the list of available options and scroll through the list to find the *Mutual Information* function. In the *Parameters: number of bins* data entry field enter a value of 8.

3. Click the *Compute* button. A matrix form containing the results matrix will appear on the desktop. This matrix has one row for each pair of columns in the data matrix. With 3 columns in the data matrix, there are 3 column pairs: columns 1 and 2, columns 1 and 3, and columns 2 and 3. The following table shows the results of the analysis:

Results of Mutual Information Analysis

Data matrix columns	Mutual Information (bits)
1 and 2	0.40197
1 and 3	0.06509
2 and 3	0.08567

The value of mutual information for columns 1 and 2 is relatively high, reflecting the contents of these columns; both columns contain 16 cycles of a sine wave with added pseudo-random noise. The noise has been added to simulate real data that would typically contain the signal of interest along with a noise component. The values of mutual information for column pairs 1 and 3, and 2 and 3, are relatively low, reflecting the fact that in both cases we are computing the value of mutual information for a noisy sine-wave and pure noise. Since the characteristics of the pure noise cannot, in principle, be predicted on the basis of the sine wave and, conversely, the behavior or the sine wave cannot be predicted on the basis of the noise, the mutual information for these column pairs is correspondingly low.

Question:

1. Why is the value of mutual information between columns 1 and 3, and between 2 and 3 not exactly 0? This might be expected given that column 3 contains pseudo-random noise.

Answer:

1. These values of mutual information would tend towards 0 as the number of data points in the column pairs increased. That is, with a sufficiently large number of data points for both the noisy sine wave and for the pseudo-random noise, the corre-

sponding value of mutual information could be made as small as we want (assuming, of course, that the pseudo-random noise component of the noisy sine waves was un-correlated with the pseudo-random noise in column 3). With a limited number of data points, there is some probability that the behaviors of the noise in column 3, the 'pure noise' column, and the noisy sine waves in columns 1 and 2 will by chance be to some small extent mutually predictable. Even with a large amount of data how-ever, there is also the chance that the noise components in the three columns will themselves be mutually predictable. This would happen because the noise is not truly random noise, but rather what is being referred to here as pseudo-random noise. Such noise is generated by an algorithm. Because of inherent limitations in any such noise algorithm, over a large enough sample of such data points, regularities will eventually appear.

References

Fraser, A. M. and Swinney, H. L. (1986). Independent coordinates for strange attractors from mutual information. *Physical Review A 33(2)*: 1130–1140.

Gray, R. M. (1990). *Entropy and Information Theory*. New York: Springer-Verlag.

Shannon, Claude. E. (1948). A mathematical theory of communication. *Bell System Technical Journal 27*: 379-423.

Correlation and Covariance Analysis

Correlation

Correlation is a measure of the degree of linear association between two variables. Cor-relation indexes the degree to which changes in the value of one variable are associated with changes in the other variable, irrespective of the variance of either of the variables. Thus, the degree of association estimated by correlation does not depend on differences in the sizes of the changes of the two variables. Correlation is positive when the variables have a tendency to increase or decrease together. The two sets of numbers [2, 4, 1, 5, 3] and [4, 7, 2, 8, 6] are positively correlated. By examining corresponding values in the two sets, we can see that they tend to increase and decrease together. Note that the size of the changes in the second set of values is greater than the size of the changes in the first set. The more that two sets of numbers tend to vary together, the higher the value of cor-relation. Correlation is negative when two variables have a tendency to vary in opposite directions. The 2 sets of numbers [2, 4, 1, 5, 3] and [2, 1, 4, 1, 2] are negatively corre-lated; by examining corresponding values in these sets, we can see that they tend to in-crease and decrease in opposite directions. The more closely that 2 sets of numbers tend to vary, the higher the value of correlation, to a maximum value of ±1. Correlation is the association measure of choice when there is reason to believe that only the pattern of changes of two sets of data is relevant and not the magnitudes of the changes themselves.

The following procedure is used to compute the value of correlation between two data sets. First, the means are removed from each of the data sets. This process is referred to as centering the data. Next, the following three quantities are computed for the centered data: (1) the sum of cross products Σxy (the sum of the products of corresponding pairs of values in the two sets); (2) the sum of squares of each individual data set, Σx^2 and Σy^2 (for each data set, the sum of the squares of each of the values in that data set); and (3) using these computed values, correlation, symbolized by r, is computed as

$$r = \Sigma xy / (\sqrt{\Sigma x^2} \cdot \sqrt{\Sigma y^2})$$

Applying this procedure to the first pair of sets of data points above, [2, 4, 1, 5, 3] and [4, 7, 2, 8, 6], the correlation between them is

$$r = 96 / (\sqrt{55} \cdot \sqrt{169}) = 0.9957$$

The correlation between these 2 data points is positive, and is in fact close to the maximum of 1.0. This finding confirms what we observe by inspecting the two sets of data points, that they vary together closely.

In data analytic practice, one application for this function is as a data-preprocessing technique that can precede neural network analysis. In some cases it might be useful, in terms of speeding network training, to preprocess the data by computing correlations. This would be done in particular if there was some reason to believe that the patterns that the network was expected to find were related to correlations. If this were the case, then the network learning task would be made easier by training the network on the correlations in the data rather than on the raw data.

Covariance

Covariance, like correlation, is a measure of the degree of association between two variables, the degree to which changes in the value of one variable are associated with changes in the other variable. The difference is that, unlike correlation, covariance does depend on the variance of the two variables. That is, the value of covariance between two sets of data depends not only on the extent to which the two sets vary together, but also on the size of the variation.

Covariance would be the measure of choice when there is reason to believe that relevant information is contained not only in the pattern of variation of two sets of data, but also in the relative magnitude of the variations. In other words, we would use covariance rather than correlation if the value of association was intended to reflect both the patterns of changes of the two data sets and the sizes of those changes.

As in the case of the correlation matrix, one application for this function is to preprocess data for neural network analysis. Covariance would be used for this purpose if there was reason to believe that the relationships the network was expected to find were related to between-variable covariances. As in the case of correlation, the network learning task would be made easier by training the network on the covariances in the data rather than on the raw data.

The Research Question

What is the degree of linear association estimated by correlation if differences in variance are not to be considered, or by covariance if variance is to be considered, between two sets of data points?

Simulnet Exercise: Creating a Correlation Matrix

In this exercise the correlation matrix will be computed for a data matrix. The elements of the correlation matrix are the correlations between pairs of data matrix columns. An element r_{ij} of the correlation matrix is the correlation between columns i and j of the data matrix. Thus, diagonal elements of the correlation matrix are one since they represent the correlation between a data matrix column and itself. The correlation matrix is symmetrical about its diagonal: Element r_{ij}, the correlation between data matrix columns i and j, has the same value as element r_{ji}, the correlation between columns j and i. The data matrix will be symbolized by X, and the correlation matrix by R.

Procedure

1. Close all forms on the desktop. Open file *intro1.dat* and minimize the matrix form. The matrix contains 3 columns of 512 rows. This matrix will be referred to as matrix X.

2. From the *Analyze* menu, select the *Matrix Functions* option. On the *Matrix Functions* dialog form, click on the *Operation* list-box to show the list of available options and scroll through the list to find the *Make correlation matrix* function.

3. Click the *Compute* button. A matrix form containing the correlation matrix R will appear on the desktop:

$$R = \begin{bmatrix} 1 & -0.088 & -0.684 & 0.441 & -0.174 \\ -0.088 & 1 & -0.222 & 0.035 & 0.443 \\ -0.684 & -0.222 & 1 & 0.003 & 0.182 \\ 0.441 & 0.035 & 0.003 & 1 & 0.599 \\ -0.174 & 0.443 & 0.182 & 0.599 & 1 \end{bmatrix}$$

This matrix contains the correlations between each pair of columns in the original data matrix. An element z_{ij} represents the correlation between columns i and j. Thus, element $r_{ij} = -0.684$. The correlation between columns 1 and 3 of the data matrix is therefore -0.684.

Questions:

1. Compare the correlation between the body temperatures of subjects 1 and 2 with that of subjects 2 and 5. For which of these pairs is temperature more highly associated?

2. Over all pairs of subjects, for which pair is body temperature most highly associated?

3. Over all pairs of subjects, for which pair is body temperature least highly associated?

Answers:

1. subjects 2 and 5.

2. subjects 1 and 3.

3. subjects 3 and 4.

Simulnet Exercise: Computing Cross-Correlations

The cross-correlation function computes the correlation between two sets of data points at some value of lag between the two data sets. To understand the meaning of lag, imagine the two data sets are each graphed on a separate strip of paper. Imagine next that the two strips are lined up, with one strip below the other. This situation corresponds to a value of lag of zero. We could now compute the correlation between the 2 time series by taking time-point 1 on the top strip, together with time-point 1 on the bottom strip, and so on for all the data points on each strip. This would be the correlation at a lag of zero. Imagine now that the lower strip is moved to the left by 10 time-points relative to the top strip. This now corresponds to a lag of 10. Again correlation is computed, now taking time-point 1 on the top time series together with time-point 10 on the bottom time series, and so on. In this situation, correlation could not be computed using all of the time-points in the top time series, since the lower time series would be short by 10 points. For moderate values of lag, up to about one-quarter of the total number of time-points in the time series, this limitation does not present a problem. For larger values of lag, some sort of wrap-around scheme would need to be implemented.

In this exercise the cross-correlation between pairs of matrix columns will be computed for values of lag between 0 and 32. Columns 1 and 2 contain 16 cycles of a sine wave with added pseudo-random noise, while column 3 contains only noise. This exercise will assume that the data in each of these columns represents an interval of 1 second, and that the data in the columns 1 and 2 therefore represent sine waves with a frequency of 16 Hz.

Procedure

1. Close all forms on the desktop. Open file *sinnois3.dat* and minimize the matrix form.

2. From the *Analyze* menu, select the *Matrix Functions* option. On the *Matrix Functions* dialog form, click on the *Operation* list-box to show the list of available options and scroll through the list to find the *Auto/cross-correlation* function. In the *Data Matrix: End column* data entry field, enter a value of 2. In the *Parameters: lag 1* data entry field enter a value of 0. In the *Parameters: lag 2* data entry field enter a value of 32. These settings will allow the cross-correlation to be computed between data matrix columns 1 and 2, for values of lag between 0 and 32.

3. Click the *Compute* button. A matrix form containing the cross-correlations matrix will appear on the desktop. Row 1 contains the cross-correlation for a lag of 0, while row 33 contains cross-correlation for a lag of 32. The first and last few rows of this results matrix is shown in the following table:

Cross-correlation: Columns 1 and 2

Lag	Correlation
0	0.64
1	0.593
2	0.594
.	.
30	0.586
31	0.643
32	0.632

These cross-correlations between data matrix columns 1 and 2 can be viewed by plotting them on an xy graph. Column 2 of the results matrix is plotted on the y axis, and column 1 of the results matrix is plotted on the x axis.

Figure 2.34 shows that correlation is a maximum for a lag of 0 and 32. With a lag of 0, the 16-Hz sine waves in the 2 columns are lined up, and we would correspondingly expect a high value of correlation. Correlation then varies in a sinusoidal manner as the two sine waves are moved past each other. A lag of 32 corresponds to a shift in 1 column past the other by 32 data points. This shift of 32 data points corresponds to 1 cycle of the sine waves in columns 1 and 2. Thus, shifting 1 time series past the other by 32 points, that is by 1 cycle, again brings the sine waves in columns 1 and 2 into alignment, giving rise to again a large value of correlation.

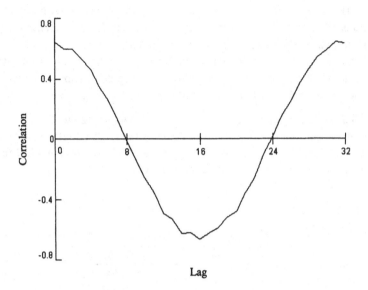

Figure 2.34. The graph shows the cross-correlation between columns 1 and 2 of matrix *sin-nois3.dat*. Since both columns contain (noisy) sine waves, the value of correlation between them varies in a sinusoidal manner as one column is moved past the other. In particular, correlation is greatest when the columns are either exactly aligned (lag = 0), aligned with column 2 one cycle over with respect to column 1 (lag = 32), or aligned with column 2 a half-cycle over with respect to column 1 (lag = 16). In this last condition, correlation has a maximum negative value because the two sine waves move in step, but in opposite directions.

4.   Click the matrix button corresponding to the original data matrix, *sinnois3.dat*, in order to set the desktop focus to this matrix. In the *Data Matrix: End column* data entry field, enter a value of 3 (it may already be 3). These settings will allow the coherence to be computed between data matrix columns 1 and 3.

5.   Click the *Compute* button. A matrix form containing the cross-correlations matrix will appear on the desktop. To view these cross-correlations between data matrix columns 1 and 3, they can be plotted on an xy graph. Column 2 of the results matrix is plotted on the y axis, and column 1 is plotted on the x axis. The graph should resemble Figure 2.35.

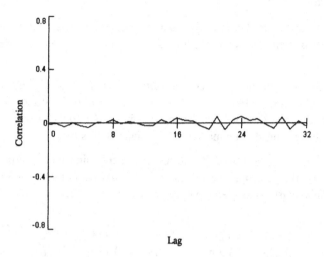

Figure 2.35. The graph shows the correlation between columns 1 and 3 of matrix *sinnois3.dat*. Since column 1 contains a noisy sine wave while column 3 contains only noise, the values of cross-correlation are low.

Questions:

1. Examining Figure 2.34 showing the cross-correlations between data matrix columns 1 and 2, why does the value of cross-correlation have a large negative value at a lag of 16?

Answers:

1. A lag of 16 represents 1/2 cycle of the 16-Hz waveform. When one of the time series data in one of the data matrix columns is moved past the other time series by one-half cycle of the sine wave, the sine waves in the two data matrix columns again line up. In this case however, when one sine wave is increasing the other is decreasing, and vice versa.

Simulnet Exercise: Creating a Covariance Matrix

This function computes the covariance matrix of the data matrix. The elements of this matrix are the covariances between pairs of data matrix columns. An element s_{ij} of the covariance matrix is the covariance between columns i and j of the data matrix. The diagonal elements of the covariance matrix are the variances of respective data matrix columns.

Procedure

1. Close all forms on the desktop. Open file *introl.dat* and minimize the matrix form. The matrix contains 3 columns of 512 rows. This matrix will be referred to as matrix **X**.

2. From the *Analyze* menu, select the *Matrix Functions* option. On the *Matrix Functions* dialog form, click on the *Operation* list-box to show the list of available options and scroll through the list to find the *Make covariance matrix* function. No other changes need to be made to the settings on this form.

3. Click the *Compute* button. A matrix form containing the covariance matrix **S** will appear on the desktop. The values in this matrix should be close to those shown below. This matrix is available preprepared as file *covar.dat*.

$$
S = \begin{bmatrix}
1.869 & -0.293 & -2.250 & 1.273 & -0.283 \\
-0.293 & 5.905 & -1.296 & 0.180 & 1.281 \\
-2.250 & -1.296 & 5.795 & 0.015 & 0.520 \\
1.273 & 0.180 & 0.015 & 4.451 & 1.504 \\
-0.283 & 1.281 & 0.520 & 1.504 & 1.418
\end{bmatrix}
$$

3
Acquiring and Conditioning Network Data

Introduction

Training a neural network involves a number of considerations in getting from the process to be modeled to the actual set of network training exemplars. These considerations include the following (Stein, 1993):

- Data Specification: Deciding on what variables should be included
- Data Collection: Collecting samples from the included variables
- Data Inspection: Inspecting the data for characteristic and anomalous features
- Data Conditioning: Preprocessing the data to extract features, correct for anomalies, or to reduce the volume of data

Data Specification

A primary consideration when a process is to be modeled using a neural network is the choice of the variables to be included in the model. There are a number of important reasons why we might not want to simply include as many variables as might be available.

1. One reason is economy. Usually, the costs involved in collecting the data cannot be ignored. More important, in the present context, are the computational costs involved in running the network. The volume of data presented to the network for training could easily be too great to allow the network to produce a model in a reasonable amount of time, or at least in the time that is available.

2. A second reason is that if variables are selected that are not relevant to the process being modeled, these variables will represent sources of noise in the data. The term noise in this context does not refer to the behavior of a stochastic process. It is meant to refer to the fact that the behavior of some variables is not correlated with the behavior of the part of the system that is being modeled. Such variables are considered to be noise variables only in the context of the task at hand, that is, to model some facet of the behavior of a system. Recall that each predictor (independent) variable that is selected becomes one of the com-

ponents of each network exemplar. Including a variable whose behavior is, in reality, not correlated with the behavior of the system being modeled will add components to each exemplar that are of no use to the network in developing the model. The presence of such extraneous variables will make it more difficult for the network to create our model, with the result that network training time is prolonged. If the number of noise variables is large, the network may be entirely prevented from training. A fundamental feature of neural network training is that the ability of a network to learn the functional relationships in the data depends crucially on the number of available training exemplars and on the amount of noise in each exemplar. To some extent, exemplars constructed by sampling the behavior of real-world variables will contain noise that is associated with the collection process itself, or that is a feature of the system being sampled. These sources of noise may not be easily controlled. Extraneous variables, however, represent a potential source of noise over which we may have some measure of control.

3.   A third reason has to do with the constraint, discussed earlier, on the ratio of the number of training exemplars to the number of network input variables (e.g., Lisboa et al., 1994). If there are more input variables than training exemplars, the classes in the data are separable by means of a linear decision boundary. More generally, the relationships in the data can be modeled using a linear function. In this situation, with more inputs than exemplars, the network can only be expected to accurately generate a linear model of the data; there are simply not enough exemplars to define a nonlinear functional relationship. With a large enough number of hidden units the network can be made to generate some sort of nonlinear model. Such a model may, however, not be one that will reliably generalize to novel exemplars.

4.   A fourth reason involves the clarity of the model that the network will generate. The smaller the number of variables selected, the simpler the resulting model becomes. While there may never be a guarantee that a system can be adequately modeled using only some small subset of the total number of system variables, such a model may at least be easier to understand, and thus provide an initial level of understanding. A model constructed on this basis may then serve as a useful first approximation to a more comprehensive model.

The only solution to the problem of choosing the right variables is to have some understanding of the process that is to be modeled. Given an appropriate set of variables, a sufficient number of exemplars, and sufficient computing power with respect to the volume of training data, we can have the reasonable expectation that the data can be successfully modeled by a neural network. For this reason it becomes vitally important to research the system to be modeled in order to identify relevant network training variables.

These points can be illustrated with an example from neurophysiology. Imagine that an experimenter measures EEG signals over some period of time and in several experimental conditions from 21 scalp-mounted electrodes. The experimenter would like to know if the resulting data can be classified according to condition. In other words, the experimenter is asking the question: Did the experimental manipulation have an effect on the recorded signals? On the basis of independent work, the experimenter hypothesizes that the relevant features of the data are contained in the signals from a subset of six electrodes. By selecting only the data from those six electrodes, the experimenter can reduce the volume of data presented to the network to about 29% of the total available data. This reduction will speed the rate of network training for the reasons identified above: First, the sheer volume of training data has been reduced. Second, if the variables, the specific electrodes, have been appropriately chosen, then the training data will be relatively less noisy. If the experimenter's hypothesis is correct, only the six chosen variables are affected by the experimental manipulation. The other 15 variables, or electrodes, represent sources of noise in the context of the experimental manipulation. The network will have that much less spurious information that it must learn to overlook.

Data Collection

Sampling and Prefiltering

Typically, processes from which data is recorded are inherently continuous in the sense that the data is available as a more-or-less smoothly changing value. The variable being sampled would correspondingly be referred to as a continuous variable. In order for a data collection system to be able to record the data in a form that can be stored in computer-readable form, and ultimately in a computer storage medium such as a disk, the continuous data stream must be converted into discrete values. These discrete values are obtained by sampling from the continuous process usually at a regular time interval. This process is referred to as analog-to-digital conversion. Of utmost importance is the rate at which the computer samples the data. This parameter is known as the sampling rate, and is simply the number of times in 1 second that the data collection system takes a measurement of, or samples, the process being recorded. Too low a sampling rate means that important features of the process may be missed if they occur between successive samples. On the other hand, too high a sampling rate means that not only is a large volume of data collected, which in itself might pose a problem in terms of storage and manipulation. More importantly, however, successive samples might now represent correlated information. Thus, if changes in the process being sampled occur relatively slowly, then sampling that process too often means that any two successive samples carry almost the same information. In order to determine what sampling rate is appropriate in any particular application, we need to have at least an approximate idea of how fast the process being recorded is changing. In alternative terms, we should decide on a value for the shortest time scale, or highest frequency, in the process that is to be recorded.

It might be decided for instance, if we were making EEG measurements, that the highest frequency component in the data that is of interest to us is 50 Hz. In that case, the data would be sampled at a rate that would capture all frequency components up to 50 Hz. What sample rate should be used to accomplish this? The answer to this question is given by the following relationship. The maximum frequency f_c that is captured by sampling every Δ seconds is given by the equation

$$f_c = 1 / (2\Delta) \hspace{4cm} 1$$

where f_c is called the Nyquist critical frequency. Equation 1 can be rewritten in terms of critical frequency f_c and sampling rate S:

$$S = 2 \cdot f_c \hspace{4.5cm} 2$$

Equation 2 translates into the criterion that in order for a sine wave to be recorded it must be sampled at least twice in one cycle. To be conservative and ensure that the highest frequency of interest is well represented in the recording, a sampling rate of up to twice this value should be used. Thus, in order to record all frequency components up to 50 Hz, we would sample at no less than 100 samples per second, and at 200 samples per second to be conservative. Reasonable values for sampling rate for this example would then be anywhere in the range of 100 to 200 samples per second.

The issue of sampling rate raises two points that are important for the data collector to keep in mind.

Over-Sampling

The first point is that a mathematical theorem exists, known as the sampling theorem, which states that if the signal being recorded has been low-pass filtered at some frequency f, then there is absolutely no need to sample the signal at a rate greater than 2f samples per second. To give an example: If the signal has been passed through a low-pass amplifier set at 75 Hz, then the signal will be completely captured with a sampling rate of 150 (75 x 2) samples per second. Sampling at any higher rate, over-sampling, will add nothing to the data except excess data points that carry little or no independent information.

Aliasing

A second point to be considered is that if, on the other hand, the signal has not been low-pass filtered or if the low-pass filter setting is numerically greater than twice the sampling rate, then power associated with frequency components above the Nyquist critical frequency f_c will falsely appear as frequency components below the critical frequency. These falsely translated, or aliased, components then represent noise in the data since their frequencies are false and the information they carry is therefore invalid.

To give an example: Suppose a signal has been low-pass filtered at 75 Hz. The appropriate sampling rate, according to the equation 2, is a minimum of 150 samples per second. Let's suppose that the signal is actually sampled by the data collection system at a rate of only 90 samples per second. The critical frequency for this sampling rate is (again using equation 2) 45 Hz. According to the sampling theorem, if the signal contains frequency components between the critical frequency, 45 Hz, and the filter setting, 75 Hz, the power associated with these components will be 'aliased', or falsely translated, into components that appear to have frequencies below 45 Hz. These noise components can then interfere with any analysis that is carried out on the sampled data.

As a second example, suppose that the sampling rate is fixed at 100 samples per second. The appropriate corresponding low-pass filter setting is 50 Hz, or lower. If the low-pass filter setting is greater than 50 Hz then the possibility exists for the power from frequency components above 50 Hz to appear as frequency components (aliases) below 50 Hz.

Analog-to-Digital Conversion

Suppose that a signal is low-pass filtered at 64 Hz as it is being recorded. The sampling rate could then be set at 128 samples per second. In other words, for every second that the data was being recorded, the data collection system measures the incoming signal level 128 times. If data is recorded for 10 seconds, the result will be 1,280 (128 x 10) samples, or time-points, for each variable being measured. The phrase time-point is sometimes used to refer to data samples when recordings are made over time. After acquiring an individual measurement, or data point, the data collection system converts the data point into a computer-readable number and then stores that number in the working (RAM) memory of the computer, or on disk.

The mechanism by which the data collection system converts a measurement into a number that can be stored in memory or on disk is known as analog-to-digital (A to D) conversion. Measurements made on continuous variables are referred to as being analog in nature. In this context the term analog simply denotes that the values recorded from the variable are, by the nature of the variable, continuous and can be represented by real numbers. Virtually all data collection, and all data analysis, systems involve digital computers. Such analog measurements must therefore be converted into a digital format that can be handled by a computer. This conversion process essentially involves converting an infinite precision real number into a necessarily limited precision binary number. This binary number consists of a string of binary digits (bits), 0's and 1's. The length of this string, the number of bits, determines the precision with which the original real number is represented. For instance, a binary number consisting of 8 bits can represent 2^8, or 256, different values. A binary number consisting of 16 bits can represent 2^{16}, or 65,536, different values. The greater the number of digits in the binary representation of the real number, the more precisely the real number can be represented by the binary number.

Different data collection systems may use different numbers of bits in the A to D conversion process. It is important that the data collector be aware of this number. If, for instance, the data collection system uses only eight-bit A to D conversion, the converted data is only accurate to about three significant figures. With 16-bit conversion the accuracy is approximately 5 significant figures. What could be potentially misleading is that when such values are stored as computer files the numbers in the files will show approximately seven significant figures. The extra significant figures are generated by the software that is used to store the numbers that come from the A to D converter onto disk (Figure 3.1).

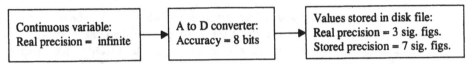

Figure 3.1. A continuous variable with infinite precision (neglecting quantum effects) is sampled by the eight-bit analog-to-digital converter to produce values that have a precision of approximately three significant figures. These values are nevertheless typically stored on disk as numbers with seven significant figures.

The argument can be made that one extra significant figure is required in the stored representation of the measured values in order to safeguard the accuracy of the stored values. Maintaining more than one additional significant figure over the number of figures that correspond to the accuracy of the A to D converter, however, can be a waste of disk storage space. For small files this extra storage space may not present a problem. For large files, or large numbers of files, the wasted space may be considerable.

If the values in a disk file consist of more than the warranted number of significant figures, the values can be rounded to the appropriate number of significant figures. If a file was created by a data collection system using an eight-bit A to D converter, for example, the appropriate number of significant figures in the file is four (three corresponding to the eight-bit conversion process, plus one guard digit). The seven significant figure values in the disk file can be processed by a rounding operation to reduce the number of significant figures to four. This will result in a reduction in the size of the file by approximately (7 - 4) / 7, or 43%, with no *real* loss of precision with respect to the original data. The following table gives the approximate conversion accuracies for A to D converters of various numbers of bits, and the corresponding appropriate number of significant figures that need to be maintained in the data when it is stored as a disk file.

File Precision vs. A to D Converter Accuracy

A to D Accuracy (no. of bits)	Converter Precision (no. of sig. figs.)	Disk File Precision (no. of sig. figs.)
6	2	3
8	3	4
10	4	5
12	4	5
16	5	6
24	8	9

Data Formats

The real numbers that represent measurements made on a continuous variable could only be completely expressed with an infinite number of significant figures. Recording such values would correspondingly require an infinite storage capacity. The practical solution is to simply truncate the representation of the real number. Numbers stored in computer-readable format may be stored in what is referred to as single precision, corresponding to approximately 7 significant figures, or in double precision, with about 15 significant figures.

Single-precision format is typically used when computational speed is more important than precision of representation. Single-precision numbers are stored within a computer using the IEEE 32-bit floating-point format, and can range from -3.40×10^{38} to -1.40×10^{-45} for negative values and from 1.40×10^{-45} to 3.40×10^{38} for positive values. Each number consists of three parts: 1 bit for the sign, 8 bits for the exponent, and the remaining 23 bits for the mantissa. These 23 bits can represent approximately 7 significant figures. Double-precision numbers are used when precision is of prime importance. Double-precision numbers are stored using the IEEE 64-bit floating-point format, and can range in value from -1.80×10^{308} to -4.94×10^{-324} for negative values and from 4.94×10^{-324} to 1.80×10^{308} for positive values. Each number consists of 3 parts: 1 bit for the sign, 11 bits for the exponent, and the remaining 52 bits for the mantissa. These 52 bits can represent approximately 15 significant figures.

Simulnet stores numbers on disk in single-precision format. Computations are performed using either double or single precision. Double precision is used whenever precision is the governing consideration. Single precision is used when the accuracy of the double-precision format is not required, and where speed of execution is more important.

Data Inspection

The first class of operation that should generally be carried out on experimental data is inspection, or visualization: graphing the data in various ways. Data visualization is one way in which the experimenter can examine the data in order to determine whether it contains features that may not be directly related to the experimental manipulation, but are, rather, artifacts of the data collection process. A decision can then be made as to whether the data containing such artifacts is to be excluded from further analysis. Having examined the data graphically, we can more easily decide what further steps should be taken before the data is subjected to more detailed analysis. One reason for preprocessing data is then to identify features such as artifacts that are to be excluded from subsequent analysis.

Graphing data is important for a number of reasons. The graphs may immediately reveal irregularities such as artifacts or outliers. Decisions can then be made about how to handle these features. Plotting the data may reveal directly that the experimental manipulation has had a greater effect on one part of the data than on another. Attention can then be focused on those parts. Using the example of EEG data recorded from multiple electrode sites, it might be possible to see directly that an effect of the experiment appears to be present to a greater degree in some channels than in others.

Data visualization functions in Simulnet include two-dimensional plots such as x-y plots, bar graphs, and polar graphs. More elaborate plotting options include three-dimensional graphs such as xyz plots and surface-contour plots that can be rotated about any axis. The following exercises will describe the creation of an xy plot and an xyz plot.

Simulnet Exercise: Creating an XY Graph

An xy graph is a graph in which one variable, often the independent variable, is plotted on the x axis of the graph, and the other variable, the dependent variable, is plotted on the y axis. In this exercise the contents of file *sincosn1.dat* will be plotted on an xy graph.

Procedure

1. Close all forms on the desktop. Open file *sincosn1.dat* and minimize the matrix form.

2. From the *Graph* menu, select the *XY Graph* option. On the *XY Graph* dialog form, make the following changes. In the *Data matrix: End column* field, enter a value of 1.

3. Click the *New Graph* button. A graph form will appear containing the graph of column 1 of the matrix. The graph should resemble the graph shown in Figure 3.2.

4. If desired, this graph can be saved to disk. To save the graph, click on the title bar of the graph form to set the desktop focus to the graph. Then, select the *Save as* option from the *File* menu. The *Save Graph* form will appear. Choose a path and file name for the graph, and then click the *OK* button.

5. Experiment with different options and settings to become familiar with this graphing function.

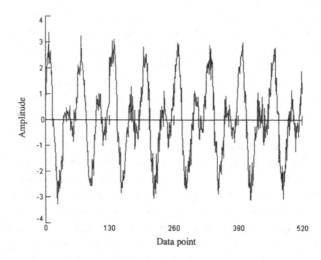

Figure 3.2. The graph shows column 1 from matrix *sincosn1.dat*, consisting of a 16-Hz sine wave added to an 8-Hz cosine wave, with added pseudo-random noise.

Simulnet Exercise: Creating an XYZ Graph

An xyz graph is a plot of three variables, one each on the x, y, and z axes. This graph can be plotted against a background of a number of different axis styles. Importantly, the graph can be rotated about any axis so that the relationship between the plotted variables can be more easily seen. Graphs can be colored using the values of any column in the data matrix, or using the velocity of the trajectory of the graph line. Velocity is computed by taking the Euclidean distance between a pair of successive data points on the graph.

In this exercise the contents of file *dufholm.dat* will be plotted on an xyz plot. This file contains data generated by integrating the Duffing-Holmes equations. This is a system of three coupled differential equations:

$$dx/dt = y$$
$$dy/dt = -dy + (x - x^3) / 2 + A \cos (z)$$
$$dz/dt = \omega$$

This system of equations is intended to model a circuit containing a nonlinear inductance driven by an AC source with a frequency ω and amplitude A. Parameter d represents the circuit damping.

Procedure

1. Close all forms on the desktop. Open file *dufholm.dat* and minimize the matrix form.

2. From the *Graph* menu, select the *XYZ Graph* option. On the *XYZ Graph* dialog form, select the *Color source: by velocity* option. For *Rotate: Angle* enter a value of 5.

3. Click the *New Graph* button. A graph form will appear containing the graph of column 1 along the x axis, column 2 along the y axis, and column 3 along the z axis. The graph should resemble Figure 3.3 (the graph in the figure has been rotated about the x and y axes).

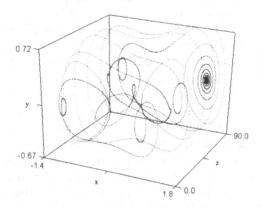

Figure 3.3. The graph shows the matrix contained in file *dufholm.dat*. The file contains 3 columns of 1,024 points, generated by integrating the Duffing-Holmes equations with an integrator step size of 0.2.

4. Click the *Rotate Left* button ◀. The graph should rotate to the left in increments of five degrees. To stop the rotation, click the same *Rotate left* button once more. All four of the direction buttons are of the maintained type: the button stays down once it has been clicked, and continues to rotate the graph until it is clicked once again.

    To reset the graph to its initial state, click the *Rotate To Center* button ◆. Experiment with the rotation buttons, using different values for *Angle*.

5. Experiment with different options and settings to become familiar with this graphing function.

Data Conditioning

The term data preprocessing refers to procedures that are carried out on data with the goal of facilitating the subsequent analysis of the data. Data transformation procedures can be subsumed under the umbrella term of data conditioning. Three general reasons for wanting to preprocess or condition experimental data are:

- Correction: To correct for presumed irregularities in the data
- Feature selection: To select only certain features in the data for further analysis
- Data reduction: To reduce the volume of data supplied to the subsequent analysis

Correction

Data correction has as its goal the management of data features that have been decided to be anomalies within the data. An initial visual examination of the data might lead to the decision to condition the data to remove some of the anomalies. As a typical example, the experimenter may decide to exclude some frequency component or range of components from further analysis. This step might be appropriate, for instance, if the data are contaminated by signals related to utility power. Such 60 Hz (or in some regions 50 Hz) interference typically occurs when low-level electrical measurements are made using high-gain amplifiers. As another example, in the particular case of EEG or MEG recordings, the experimenter may decide at the start of the experiment to exclude from analysis portions of the record that include muscle-movement artifacts such as eye-blinks.

We may also decide to transform, or condition, data in order to remove features that may or may not be artifacts of the data collection process, but that nevertheless may adversely affect the course of any subsequent analysis. One example of such a feature is the presence in the data of outliers, relatively rare and extreme values. If there is independent reason to believe that such outliers are atypical of the data, these values can be removed in order to avoid the unwarranted prejudicial effect they may have on any subsequent analysis. Since by definition the values of outlying data points are large compared with the mean of the data, such data points can have a disproportionate effect on later analyses, possibly obscuring more subtle features in the data.

Still another situation in which we might want to transform the data prior to analysis, particularly when that analysis involves neural networks, is to correct for severe nonnormality of the data distribution. In the case of analytic procedures such as neural network classification, the task of the network is generally made somewhat easier if the network training data is at least approximately normal in distribution. A variety of transformations are available to correct for this feature.

Feature Selection

A second reason for conditioning data is feature selection, to extract a subset of the features within the data and to present only that subset to the subsequent analysis.

One motive for feature selection is to reduce the ambiguity in the results of subsequent analysis, again particularly when neural networks are involved. This is so because the network is asked to construct a model using only a subset of the available data features. Such a model may be correspondingly easier to interpret than a model that involves a greater number of data features. In the limit, if the neural network is presented with data containing only a single feature, such as a single frequency component, then the resulting model can confidently be expected to be based on only that feature. With an increasing number of data features presented to the network, it becomes increasingly difficult to judge how the resulting model makes use of the different features.

A second motive for feature selection is to attempt to correct for deficiencies in the data specification process. For instance, network training data may include correlated variables. Variables that are correlated with other variables add no new information to that supplied by the others. On the other hand, the correlated variables represent additional data that the network will have to contend with in constructing its model. Network training will, as a consequence, be slowed down. Such correlated variables can be effectively removed by techniques such as the principal component analysis.

As an example of data conditioning for feature selection, the data could be filtered to extract for analysis some range of frequency components. This might be the case, for instance, if the hypothesis predicted an effect of the experimental manipulation on a particular range of frequencies preferentially. For instance, an experiment might be conducted in which EEG recordings were made in both an eyes-open and an eyes-closed condition. It might be decided beforehand, on the basis of pilot studies or a literature survey, that a pronounced effect of this manipulation was expected in the 8- to 12-Hz range of frequencies in the EEG record. Correspondingly, the experimenter might decide to filter the raw EEG data to extract this particular frequency range, and then subject only these frequencies to neural network analysis. If the network is able to adequately classify the data according to experimental condition, then the experimenter could conclude that there is support for the hypothesis that the experimental manipulation has had an effect on the particular range of frequencies that were selected. Had the network instead been supplied with all of the recorded frequency components, successful classification would only have indicated that the experiment has had an effect on some unknown features in the data The experimenter would not however know if the affected data feature was some subset of frequency components.

As a second example, suppose that the experimenter intended to train a neural network to classify EEG data recorded in an experiment with two conditions. The experimenter might entertain the hypothesis that the experimental manipulation should have an effect on the EEG in terms of the pattern of between-channel correlations. Given such a hypothesis, it would be appropriate to preprocess the data by computing between-channel correlations. The neural network would then be trained on these correlations, rather than being trained on the raw EEG data. If the neural network was able to distinguish between data from the different experimental conditions on the basis of the correlations, the experimenter could conclude that there was support for the initial hypothesis. The experiment has produced an effect on between-channel correlations. As an alternative, suppose the network was trained on the raw EEG data, and again the network was able to distinguish between the data from the different experimental conditions. In this situation there would be no way of knowing just what features in the data accounted for the successful discrimination performance of the network.

Data Reduction

A third motive for carrying out data preprocessing prior to neural network analysis is data reduction. Reducing the volume of data, in a way that does not compromise the relevant information content, can lessen the computational load on a neural network. The result would be a corresponding reduction in the time required to carry out network training. This may not only be convenient but indeed essential if the analysis is to be carried out within a fixed time frame. Neural networks are inherently parallel processes, but they can easily become effectively unusable when implemented on a serial machine if, at the same time, they are asked to deal with a large volume of data. This issue is of course related to the level of available computing power. Data, like work, seems always to expand to fill the resources available to deal with it.

Often the two goals of data preprocessing, feature selection and data reduction, are not independently achieved. In the example of EEG data analysis, the raw EEG data would typically consist of single-trial recordings from 21 channels with, say, 1,000 data points for each channel. The total number of data points in 1 such trial would then be 21 x 1,000, or 21,000 data points. Typically many such trials would be recorded in each of the experimental conditions, with each trial forming a single-network exemplar. Each exemplar would then contain 21,000 predictor variables—each data point would in this situation be considered to be a single variable, of which there would be multiple cases, one from each trial. This number of predictor variables is at once too large to allow training to be carried out in a reasonable amount of time, and too large to allow the neural network to develop a credible nonlinear decision boundary between the classes (the experimental conditions that the network is to learn to distinguish). For 21 channels of data, the number of correlations is (21) x (20) / 2, or 210. If the network was trained on only these correlations, each network exemplar would consist of only these 210 correlations as the predictor variables. The number of predictor variables has been reduced from 21,000 to 210. Of course the network is now being trained on only the between-channel correlations. A successful network training outcome would support the initial hypothesis. An unsuccessful outcome, that is if the network is not able to distinguish between the classes of trials, would mean that the initial hypothesis should be reexamined. At this point the experimenter could also decide to try a more exploratory approach, preprocessing the data to extract still other features on which to train the network.

In summary, data transformation is typically applied to data prior to analysis for any combination of three reasons. The first reason is *correction*: To correct for features in the data that have been decided to be anomalous. The second motive is *feature selection*: To extract subsets of the features in the data for further analysis. The third reason is *data reduction*: To reduce the computational load on a subsequent analytical procedure.

Simulnet Data Conditioning Functions

The following list of Simulnet functions describes a selection of the more commonly used transformations, and the purpose they serve in the preprocessing stage of data analysis. In most cases a transformation may serve more than one of the motives that have been discussed. The names of the transformations correspond to their names within Simulnet.

Table 3.1 Selected Data Transformations

Name	Purpose
Detrend—Order 0	mean, or zero-order trend, is removed from the data.
Detrend—Order 1	slope, or first-order trend, is removed from the data.
Frequency Filter	one, or a range, of frequency components may be optionally selected or rejected.
Heaviside	transforms the data into binary values; the output is 1 if the data value is greater than 0, and 0 otherwise.
Inverse	normalizes an extremely skewed distribution.
Logarithm	normalizes a moderately skewed distribution.
Principal Component Filtering	a subset of the dimensions in the data are extracted using the Karhunen-Loeve transformation. This subset is presumed to account for most of the variance in the data, while the rejected dimensions are presumed to contain mostly noise.
Round values	rounds the values to a specified number of significant figures. This operation can be used to reduce the size of a disk file. The operation may be appropriate if the data in the disk file has been stored with a greater number of significant figures than is warranted by the number of bits of accuracy in the analog-to-digital conversion within the data collection system.
Square root	normalizes a slightly skewed distribution.

Name	Purpose
Standardize Columns; Standardize Matrix	matrix columns are standardized. Standardization involves two operations: Setting the mean to 0, and setting the standard deviation to unity. By standardizing a number of data samples, differences between them in terms of mean and standard deviation are removed. Standardizing over individual columns uses only data within each column for computing mean and standard deviation; standardizing over the matrix uses the entire matrix for these computations.
Trim Outliers	outliers, values further from the mean than some specified number of standard deviations are trimmed.

These data transformation functions are described in more detail in the following sections.

Detrend—Order 0

Introduction

The zero-order trend of a set of measured values, or data points, is the mean. As a simple example, consider the following hypothetical set of measured values [2, 4, 3, 1, -2, 5, 3, 4]. The mean of these values is 2.5. Subtracting this mean from each of the measured values, we obtain [-.5, 1.5, 0.5, -1.5, -4.5, 2.5, 0.5, 1.5]. The mean, or in alternate terms, the zero-order trend, has been removed from the original set of data values.

One reason for carrying out this transformation on a set of data points is that the mean value represents information that is not considered to be useful for immediate purposes. One reason for this might be that the mean represents some artifact of the data collection process, rather than a feature of the process that is being measured. An EEG record made over some number of seconds for example frequently has a non-zero value of mean. This non-zero value of mean is most likely generated by the EEG amplifiers, rather than by the subject. On the assumption that this is the case, it would be appropriate to remove the mean value from EEG record. This could be accomplished by calculating the mean of each EEG channel separately, and then subtracting the channel mean from the data points within each channel.

Description

The *Detrend—Order 0* function removes the mean from each column of the matrix. This function computes the mean separately for each matrix column. This column mean is then subtracted from each element within that column. This process is then repeated for each column in the matrix.

Simulnet Exercise: Detrending a Matrix

1. Clear the desktop of all forms. Open file *intro1.dat* and minimize the matrix form. This matrix will be referred to as **X**, and should contain values as shown in the following:

$$
\mathbf{X} =
\begin{bmatrix}
40.1 & 35.0 & 41.3 & 41.9 & 38.4 \\
38.8 & 40.5 & 40.7 & 37.6 & 37.1 \\
39.2 & 40.9 & 37.7 & 38.8 & 39.5 \\
37.0 & 40.1 & 42.0 & 40.6 & 39.7 \\
37.1 & 35.2 & 41.3 & 35.3 & 36.8 \\
40.6 & 38.0 & 35.3 & 39.3 & 37.0
\end{bmatrix}
$$

2. From the *Transform* menu, select the *Transformations* option. On the *Transformations* dialog form, click on the *Operation* list-box to show the list of available options, and scroll through the list to find the *Detrend - Order 0* function. No other changes need to be made to the settings on this form.

3. Click the *Compute* button. A matrix form containing the detrended data matrix **C** will appear on the desktop. The values in this matrix should match those shown below (rounded to three significant figures). This matrix is available pre-prepared as the file *intro3.dat*.

$$
\mathbf{C} =
\begin{bmatrix}
1.30 & -3.28 & 1.58 & 2.98 & 0.32 \\
0.00 & 2.22 & 0.98 & -1.32 & -0.98 \\
0.40 & 2.62 & -2.02 & -0.12 & 1.42 \\
-1.80 & 1.82 & 2.28 & 1.69 & 1.62 \\
-1.70 & -3.08 & 1.58 & -3.62 & -1.29 \\
1.80 & -0.28 & -4.42 & 0.38 & -1.08
\end{bmatrix}
$$

Standardize Columns

Introduction

This function will standardize a set of data values. The standardized values are computed as follows: First, the mean of the data values is computed. Second, the standard deviation of the values is computed. Standard deviation is, in approximate terms, a measure of how much the data values vary about their mean value. Third, each data point is standardized by first subtracting the mean from the value, and then dividing the result by the standard deviation. For the set of data values used in the preceding exercise, [2, 4, 3, 1, -2, 5, 3, 4], the standardized values (rounded to 3 significant figures) are [-.243, .728, .243, -.728, -2.18, 1.21, .243, .728].

An example may help to make clear why it is sometimes useful to standardize matrices. Suppose that each of two matrices contained the brightness values for a digitized image, a different image in each matrix. Each matrix thus contains a representation of one of the images. Suppose further that the task was to compare the images to compute a measure of their similarity. Typically, at the start the images might differ in brightness and in contrast. These differences may be important features distinguishing the two pictures. In this example, however, it will be assumed that brightness and contrast are not important to the task. The goal then is to develop a measure of similarity between the two images that is independent of differences in brightness and contrast.

It turns out that the overall brightness of an image is reflected in the mean value of the elements of the corresponding brightness matrix. Further, the overall contrast of an image is reflected in the standard deviation of the elements in the brightness matrix. If the means are removed from the values in both of the matrices, the two image representations will have been equalized for brightness. If the results in the two matrices are divided by their corresponding standard deviations, the two image representations will have been equalized for contrast. The two matrices have been standardized, and the image representations contained in the matrices have been equalized for both brightness and contrast. Any differences remaining in the two matrices should then represent the differences that are being assessed.

It is sometimes useful to be able to standardize matrices containing data of any type in order to make more salient features in the data other than mean and standard deviation.

Description

The data matrix will be denoted by **X**, and the standardized data matrix by **Z**. This function standardizes the values within each matrix column. For each column each value in the column is replaced by the corresponding standard score. For each column x_j we compute the mean, symbolized as Mean(x_j), and standard deviation, symbolized as SD(x_j). The standard score z_{ij} for an element x_{ij} of column x_j is then computed as

$$z_{ij} = [x_{ij} - \text{Mean}(x_j)] / \text{SD}(x_j)$$

That is, from each element within a column, the column mean is subtracted, and the result is divided by the column standard deviation. This process is then repeated for each column in the matrix.

Simulnet Exercise: Standardizing a Matrix

1. Clear the desktop of all forms. Open file *intro1.dat* and minimize the matrix form.

2. From the *Transform* menu, select the *Transformations* option. On the *Transformations* dialog form, click on the *Operations* list-box to show the list of available options, and scroll through the list to find the *Standardize Columns* function.

3. Click the *Compute* button. A matrix form containing the standardized data matrix **Z** will appear on the desktop. The values in this matrix should match those shown be-

low (rounded to three significant figures). This matrix is available pre-prepared as the file *intro2.dat*.

$$
Z = \begin{bmatrix}
0.95 & -1.35 & 0.66 & 1.41 & 0.27 \\
0.00 & 0.91 & 0.41 & -0.62 & -0.83 \\
0.29 & 1.08 & -0.84 & -0.055 & 1.19 \\
-1.32 & 0.75 & 0.95 & 0.80 & 1.36 \\
-1.24 & -1.27 & 0.66 & -1.71 & -1.08 \\
1.32 & -0.12 & -1.83 & 0.18 & -0.91
\end{bmatrix}
$$

Frequency Filtering

Introduction

This function allows one or more matrix columns to be filtered, using any one of a number of filter functions. Most real-world signals are complex in that they consist of more than one frequency component. As an analogy, an auditory signal consisting of a single-frequency component would be a 'pure' tone. Such tones are rare in nature. Generally, natural sounds are more complex than pure tones, and consist of many frequency components. Virtually any real-world sound falls in this category. Natural auditory systems are in fact capable of analyzing sounds into their frequency components. A person can tell, for instance, that a complex tone might consist of a high and low tone.

Suppose that the goal was to analyze only one of the components of such a complex natural sound. One approach would be to frequency filter a recording made of this sound, to retain only the one component of interest and to eliminate all others. Frequency filtering can be used to remove unwanted frequency components from a signal. Such components would be recorded along with the signal of interest, because of limitations in the recording system, or because the nature of the signal itself. Thus, for example, when making EEG or MEG recordings, the signal contains not only the physiological signal that is of interest to us, but typically also frequency components due to electrical power wiring in the vicinity of the recording equipment. The electrical supply has a frequency of 60 Hz (or 50 Hz in some areas). As a result, EEG or MEG recordings typically contain components at the frequency of the utility power supply. These components generally need to be eliminated before the signal can be used for analysis. This elimination is done by filtering the signal appropriately. Such filtering can be carried out either as the signal is being recorded (on-line filtering), or after it has been recorded and stored on disk (off-line filtering).

This exercise will use an already prepared matrix (file *sinnois2.dat*) that contains data whose characteristics are known, noisy sine waves. The matrix consists of 8 columns, each of which has 512 rows. The data in each of these columns consist of 16 cycles of a sine wave, with added pseudo-random noise. This exercise will assume that these 16 cycles represent an interval of 1 second, and that the data correspondingly represents a sine wave with a frequency of 16 Hz. In the following procedure, 1 column of this matrix will be filtered to extract only the 16 Hz component and to remove all of the added noise.

Simulnet Exercise: Band-Pass Filtering a Matrix

1. Clear the desktop of all forms. Open file *sinnois2.dat* and minimize the matrix form. This matrix will be referred to as the data matrix.

2. Graph column 1 of the matrix using the following procedure:

   - From the *Graph* menu, select the *XY Graph* option.
   - On the *XY Graph* dialog form, enter 1 in the *Y axis: End column* field.
   - Click the *New Graph* button. A new graph will be created that should resemble Figure 3.4. This graph shows 16 cycles of sine wave with added pseudo-random noise.

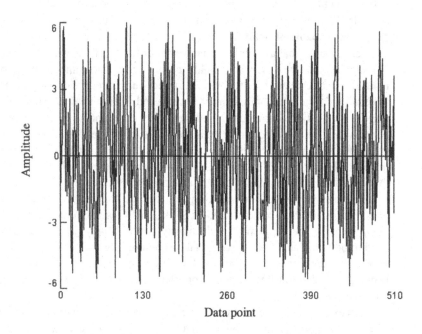

Figure 3.4. The graph shows column 1 from matrix *sinnois2.dat*, consisting of a 16 Hz sine wave, with added pseudo-random noise.

3. From the *Transform* menu, select the *Filter* option. On the *Filter* dialog form make the following changes:

    *   In the *Data matrix: End column* field, enter a value of 1.
    *   From the *Filter Type* list, select the *Band-pass* option.
    *   In the *Parameters: Lower corner frequency* field, enter the value 16.
    *   In the *Parameters: Upper corner frequency* field, enter the value 16.

4. Click the *Compute* button. A matrix form containing the results matrix *sinnoi_1.dat* will appear on the desktop. This matrix contains a single column containing the filtered data matrix column.

5. Graph matrix *sinnoi_1.dat* using the following procedure:

    *   Click the matrix button corresponding to this matrix.
    *   Click the *Redraw* button. The graph will be redrawn, and the result should look like Figure 3.5. The 16-Hz sine wave is now clearly evident.

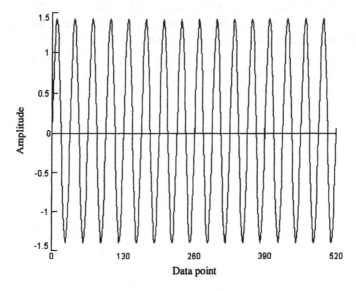

Figure 3.5. The graph shows the filtered data in matrix *sinnoi_1.dat*, consisting of 16 cycles of a sine wave.

Principal Component Analysis

Introduction

The goal of principal component analysis is to find and make use of information about the nature of the variables in a matrix. Recall that variables are assumed to be contained

within matrix columns. The data in the columns of a matrix may or may not all be mutually independent. It may be, for instance, that some columns can be expressed as linear
combinations of other columns. In some applications, such as data reduction, it can be
useful to know that the original data matrix can be re-expressed as a matrix with a
smaller number of mutually uncorrelated columns. Another area of application, filtering
out noise from a data matrix, makes use of the following fact: If the data in the original
matrix contains noise that is uncorrelated with each of the matrix variables, this noise can
be extracted from the matrix and expressed as one or more separate variables. One of
these variables can thus account for purely stochastic noise in the data. Others of these
variables can account for factors in the data that are considered to be noise because they
are unrelated to the variables of interest. These 'noise' variables can be eliminated from a
data matrix. The result will be a new matrix that is an approximation to the original data
matrix, but synthesized using the reduced number of variables.

Principal component analysis will be explained using a geometrical approach. The
data in the original data matrix, denoted by X, can be viewed as a set of data points in a
space similar to the phase space discussed earlier. This space is referred to as the variable
space of the data since it is defined by a set of dimensions, each of which corresponds to
one of the data variables. To illustrate, suppose that the data matrix contains three columns, corresponding to the three data variables. This data can be visualized as a group of
points in a three-dimensional space, as shown in Figure 3.6. This graph has been created
by considering the set of three points in each matrix row as the coordinates of a single
data point in the three-dimensional space.

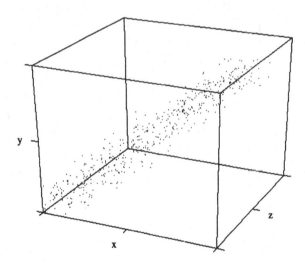

Figure 3.6. The graph shows the data in the original matrix X. The set of data points forms an approximate ellipse, with the major axis of the ellipse extending from the lower front left corner to
the upper back right corner of the plot. The major axis thus lies at an angle to the x, y, and z axes.

The goal of the principal component analysis is to define a new set of coordinate axes that fulfill two conditions. First, each of the new axes must be parallel to one of the major axes of the ellipse formed by the data points (referred to as the data ellipse). Second, the new axes must be mutually orthogonal. That is, the new axes must all be oriented at right angles to each other. The orientation of these new coordinate axes is defined by what are termed the eigenvectors of the data matrix. Each of the new axes is defined by its own eigenvector. The matrix consisting of all the eigenvectors is denoted by U. The columns of matrix U, the eigenvectors of data matrix X, prescribe how the original axes must be transformed in order to create the new set of mutually orthogonal axes. In order to rotate the data ellipse in matrix X so that the axes of the ellipse lie parallel to the new set of axes, matrix X is multiplied by matrix U. The result is a new matrix denoted by XU. Then, in order to make the values in XU independent of the variance of the columns of matrix X, XU is standardized by dividing each of its columns by the variance of that column, $D^{1/2}$. These two successive multiplications produce matrix Z:

$$Z = XUD^{-1/2}$$

This process of rotating matrix X can also be conceptualized as the creation of a set of new variables out of the original variables in matrix X. When the new set of axes are created by multiplying matrix X by the matrix of eigenvectors, a new set of variables is in effect created. Each of these variables is a linear combination of the original variables, with the constraint that the new variables must be mutually uncorrelated. Thus, each of the columns of matrix Z represents one of these new, mutually uncorrelated variables.

To illustrate this process, matrix X can be rotated to create matrix Z. Then, the columns of the matrix Z can be plotted on a three-dimensional graph. Recall that the axes of this graph will now correspond to the new variables constructed by taking linear combinations of the original data variables, subject to the two conditions discussed earlier. The axes of the data ellipse should now lie parallel to the axes of the graph. As shown in Figure 3.7, only the major axis of the data ellipse is clearly evident. The second and third axes are approximately equal. Plotting the data in matrix Z produces the graph shown in Figure 3.7. Clearly, the major axis of the data ellipse now lies parallel to the x axis.

Principal component analysis requires that at some point the data matrix be converted into a symmetric covariance or correlation matrix. If such a matrix is already available it can be passed to the principal component analysis directly. Usually, however, only the original data matrix is available. The appropriate option on the *Principal Component Analysis* dialog form can be selected to create the covariance matrix automatically when the analysis is carried out. The principal component analysis then reduces the correlation or covariance matrix to tridiagonal form, using a technique known as a Householder reduction. Next, the eigenvalues and eigenvectors of the reduced matrix are computed using the QL algorithm. Using the QL algorithm, the covariance or correlation matrix is factored to produce the matrix of eigenvalues D, and the matrix of corresponding eigenvectors U. Eigenvalues specify the relative amount of variance in the data in terms of a set of mutually uncorrelated variables that are formed by taking linear combinations of the original data variables. The way in which these linear combinations are taken is de-

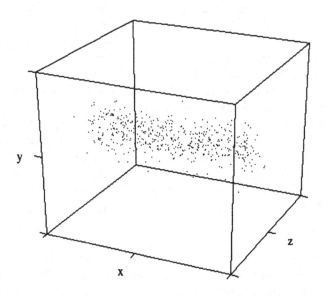

Figure 3.7. The graph shows the data in matrix **Z**, created by multiplying data matrix **X** by the eigenvector matrix **U**, and then standardizing the result. The set of data points is seen to have been rotated so that the major axis of the data ellipse lies parallel to the x axis (the plot has been scaled to place the data points in the approximate center of the graph).

fined by the eigenvectors of the data matrix. The amount of variance corresponding to each eigenvector is equal to the square root of the eigenvalue.

Using the original data matrix, and the eigenvalue and eigenvector matrices as starting points, the principal component analysis creates the following additional matrices:

- **Z** = **XUD**$^{-1/2}$, where **Z** is the matrix of standardized principal component scores. This multiplication can be broken down into two operations: First, by multiplying the original matrix **X** by the eigenvector matrix **U**, the original variables of matrix **X** are linearly combined to create a new set of variables. Another way of describing this is that the coordinate axes defining the variable space of the data are rotated, creating a new set of coordinate axes. The way in which the original variables are combined to create the new variables is defined by the eigenvectors. What is significant about these new variables, is that they are mutually uncorrelated, even if the original variables are not. In terms of the rotation of coordinate axes, whatever the relative orientation of the original axes, the new set of axes are all at right angles to each other: They are mutually orthogonal. The second part of this operation involves multiplication by **D**$^{-1/2}$, the inverse of the square root of matrix **D**. Through this multiplication the rotated matrix (**XU**) is standardized: The variance of each of the new columns is set to one by dividing each column by the square root of the corresponding eigenvalue.

- $F = UD^{1/2}$, where F is the matrix of factor loadings. Factor loadings are weights that are applied to the standardized component scores to reproduce the original data matrix.

- $X' = ZF^T$, where X' is the reproduced data matrix (the superscript T denotes the transpose). In constructing matrix X' there is a choice about how many of the new, uncorrelated variables to use. If all are used, the result is essentially the original data matrix X. If however, the number of variables used less than the total, then the new matrix X' is only an approximation to the original matrix. The choice of how many variables to use in this reconstruction can be made by noting the relative sizes of the eigenvalues. There is one eigenvalue for each variable. Eigenvalues are presented sorted in order of decreasing size. Thus, the first eigenvalue corresponds to the new variable that by itself accounts for the largest portion of the variance in the original data matrix. Similarly, the second eigenvalue corresponds to the new variable that accounts for the next largest portion of the variance, and so on. If it is found that, out of a total of n eigenvalues, only the first m are larger than one, an appropriate decision would be to use only the first m variables in constructing matrix X'. These first m eigenvalues correspond to the m new variables in matrix Z that account for the largest portion of the variance in the original data matrix. The criterion of retaining eigenvalues greater than one is based on the assumption that eigenvalues less than one account for less variance than a single original variable. If stochastic noise is evenly distributed among the original variables, then eigenvalues less than one will correspond to new variables that contain more noise than signal of interest. Other heuristics can also be applied to this process of deciding how many eigenvalues to retain in constructing the approximation to the original data matrix.

Carrying out the principal component analysis involves two stages:

1. In the first stage of the analysis, the function is asked to produce D (termed the *scree* matrix on the dialog form) the matrix of eigenvalues. For the *Number of retained components* parameter on the *Principal Component Analysis* dialog form, enter the total number of columns in the original data matrix. The scree matrix is then inspected in order to determine the number of significant eigenvalues. Denote the number of significant eigenvalues by m.

2. In the second stage of the analysis, enter m for the *number of retained components*. The function is now asked to generate other matrices depending on the goal of the analysis. If the goal is to carry out principal component filtering, the analysis is asked to generate the reproduced data matrix X. Matrix X will be an approximation of the original data matrix, using only the specified number of retained components. If the goal of the analysis is to carry out data reduction, the function is asked to generate the matrix of standardized principal component scores, denoted by Z. This matrix contains mutually uncorrelated columns that are linear combinations of the columns in the original data matrix.

Principal Component Filtering

This function will perform a filtering operation on a data matrix based on the principal components in the data. The aim of principal component filtering is to try to reduce the amount of noise in the data, where noise can be defined as random variations within the data. Such random variations are presumed, for the purpose of this filtering procedure at least, to be unrelated to the experimental manipulation or to any systematic effects of the experiment. After such noise components are removed, the resulting data will presumably represent a clearer picture of the information of interest within the data.

Principal component filtering can be related to the operation of frequency filtering. In the case of frequency filtering, the data is decomposed into a series of frequency components. Another way of looking at this is that a number of dimensions within the data are identified. These dimensions are the various frequency components. It is decided, when doing frequency filtering, that some of these dimensions, or frequency components, may be discarded and only a subset selected for further analysis. Frequency filtering thus involves selecting, generally on the basis of some initial hypothesis, a subset of the dimensions (the frequency components) within the data.

In the case of principal component filtering, the data is similarly decomposed into some set of dimensions referred to as principal components. These components are not, however, selected on the basis of frequency. They are selected on the basis of how much of the variance within the data can be accounted for by each of these dimensions. Commonly, the procedure used to identify these dimensions first 'repackages' the variance within the original data into a sequence of components. The first component is one that by itself can account for the largest amount of variance. The second component is chosen to account for the next largest amount of variance with the restriction that this second component must be independent of (orthogonal to) the first principal component, and so on for the third and higher components. Each new component, accounting for less and less of the variance in the original data, is smaller than the last. Eventually, the variance that is being accounted for by these ever smaller components is presumed to be due to a stochastic component, that is random noise, in the data. Such noise components can then be eliminated from the data, much as unwanted frequency components are eliminated in the frequency filtering operation.

As in frequency filtering, assumptions have to be made regarding which components are to be retained and which components are to be discarded. In the case of frequency filtering, the experimental hypothesis may be a guide as to which frequency components are to be kept. In the case of principal component filtering, the assumption that is made is that random noise is evenly distributed among the various principal components. Variance due to the experimental manipulation, however, is assumed, not always validly, to be distributed so that the largest chunk resides in the first, largest frequency component, with successively smaller amounts being contained within successive components. When the size of the component is equal to the noise variance in one of the original data variables, it might be considered to primarily represent noise. In other words, the largest component has the largest signal-to-noise ratio, as it were. Successive components represent successively smaller signal-to-noise rations. When the signal-to-noise ratio is less than one, the component is typically rejected.

The amount of variance associated with each of the frequency components is indicated by the magnitude of one of the eigenvalues of the data. Eigenvalues are numbers that are generated by the principal component analysis. There is one eigenvalue for each column, and thus for each variable, in the original data matrix. Recall that the total variance in the original data matrix is repackaged into new bundles, each bundle corresponding to one of the principal components. Each principal component has associated with it an eigenvalue. The magnitude of the eigenvalue is a measure of how much of the original variance is accounted for by that eigenvalue.

If the original data matrix of n columns contained only random noise, the n eigenvalues from the principal component filtering operation would all be of equal value, and the value would be one. An eigenvalue with a magnitude of one is therefore assumed to be associated with a principal component that contains as much noise as signal. The term signal in this context refers to information in the data matrix not due to random variation. Eigenvalues with magnitudes less than one are therefore assumed to represent components with more noise than signal. In sum, by examining the relative sizes of the eigenvalues from the principal component analysis, we can estimate how many of these components contain mostly signal and how many contain mostly noise. With a data matrix of n columns of experimental data, it is often found that only some number m, less than n, of these eigenvalues have magnitudes greater than one.

In sum, the columns in the data matrix contain measurements of some number of variables. The assumption is made that the data collection process has accumulated random noise along with the signal of interest. The goal is to carry out principal component filtering on this data matrix in order to try to remove as much of this noise as possible, while retaining as much of the signal as possible.

The principal component analysis procedure involves two stages. In the first stage, the analysis provides a list of the eigenvalues. Examining this list, and applying some heuristic, a subset of the eigenvalues is selected. In the second stage, the analysis is asked to use only the selected number of eigenvalues to construct a filtered version of the original data matrix.

Stage 1: Deciding How Many Eigenvalues To Retain

The total number of eigenvalues provided by the principal component analysis will be equal to the number of columns in the original data matrix, which is also equal to the number of new, composite variables created by the function. The size of each of these eigenvalues corresponds to the amount of variance in the original data matrix that is accounted for by each of these composite variables. The principal component analysis will present the eigenvalues sorted in order of decreasing size. The first, largest eigenvalue corresponds to the new composite variable that by itself accounts for the largest portion of the variance in the original data matrix. The second largest eigenvalue corresponds to the new variable that accounts for the next largest portion of the variance, and so on. Of this total, a subset of the largest eigenvalues is selected. If, for example, the original data matrix contains 24 variables, the analysis will provide a list of 24 eigenvalues. On the basis of some heuristic such as keeping only eigenvalues greater than one, it may be decided that only the largest 12 of these are significant. The corresponding 12 new variables that would be created by the analysis would account for the largest portion of the variance in the original data matrix. The 12 eigenvalues that were not retained represent variance that is mostly due to noise. In some cases, it may not be easy to make this decision about how many eigenvalues to retain. As well, the principal component analysis may take a significant amount of time to carry out if there are many predictor columns.

How do we decide how many eigenvalues to keep? The answer is generally not straightforward. Examining the list of eigenvalues reveals that, while the size of the eigenvalues decreases steadily, it almost never drops to zero with real-world, noisy data. Here, noise refers to random variations in the data due to the data collection process, or a random component to the behavior of the system under study. Given that such noise is evenly distributed among the measured variables, the data in each of the data matrix columns can be thought of as being composed of two components. One component is the record of the behavior of the column variable. The second is the random noise. As a consequence, the magnitude of each of the eigenvalues will be comprised of two parts, one of which corresponds to the variance of the variable itself, and one that corresponds to variance of the noise. Since all columns are assumed to contain approximately the same amount of noise, all eigenvalues will contain an approximately equal noise component. This noise component will result in a non-zero value for even the smallest of the eigenvalues. It is generally assumed that large eigenvalues correspond to combinations of the original variables that account for variance due mostly to the behavior of the variables, rather than to noise. Small eigenvalues are assumed to correspond to combinations of the original variables that account for mainly noise variance. In sum, a criterion is needed for discriminating between these two situations: eigenvalues representing variance due to the system variables, and eigenvalues representing variance due to noise.

Eigenvalues can be selected on the basis of a number of heuristics. One of these is that eigenvalues should have magnitudes greater than one. The rationale for this criterion is that an eigenvalue with a size less than one can be shown to represent a combination of original variables that accounts for less variance than the variance of one of these variables. If noise is evenly distributed among the original variables, such an eigenvalue must therefore represent more noise variance than signal variance. A second heuristic that can be applied in deciding how many eigenvalues to keep is that eigenvalue pairs should not be split. Eigenvalues are considered to form a pair when their magnitudes are relatively similar in comparison with the magnitudes of neighboring eigenvalues.

Stage 2: Generating the Standardized Component Scores Matrix

The second stage of the procedure is now carried out; the principal component analysis is run a second time. This time, however, the analysis is asked to use only the number of eigenvalues that we have decided to retain. As well, in this second run, the analysis is asked to generate the standardized component scores matrix Z. Each column in this matrix represents a linear combination of the original columns. That is, each variable represented by the columns of matrix Z is a linear combination of the variables in the original matrix. These new variables are designed to be mutually uncorrelated. By retaining only such linear combinations corresponding to some subset of the largest eigenvalues, it is assumed, and it is unfortunately only an assumption, that the effect of noisy and/or correlated variables in the original data matrix has been removed. A potential danger is that the effects of interest in the data may account for only a small proportion of the total variance in the data. In such an event information related to these small effects would be lost along with some of the smaller eigenvalues. There is unfortunately no easy solution to this potential problem.

Simulnet Exercise: Principal Component Filtering

In this exercise a principal component filtering operation will be carried out on the matrix contained in file *sinnois1.dat*.

Procedure

1. Close all forms on the desktop. Open file *sinnois1.dat* and minimize the matrix form. This matrix will be referred to as the data matrix.

2. From the *Analyze* menu, select the *Principal Component Analysis* option. On the *Principal Component Analysis* dialog form, select the *Output: e values* option. Deselect all other options in the *Output* group. Select the *Computations: Make covariance matrix* option.

3. Click the *Compute* button. After a few seconds, a matrix form containing the eigenvalues matrix will appear on the desktop. This matrix will be called *e_values.dat*. The values in this matrix should be close to those shown in the following table (rounded to three significant figures). This matrix will contain one column and eight rows. Each row contains the magnitude of one of the eigenvalues of the data matrix. Notice that only the first eigenvalue, in row 1, is larger than 1. All other eigenvalues have magnitudes of less than one. After inspecting matrix *e_values.dat*, close the matrix form to remove it from the desktop.

Eigenvalue
8.30
0.415
0.381
0.343
0.336
0.332
0.310
0.269

4. Enter a value of 1 in the *Computations: No. of components* data entry field. Make sure that the *Computations: Make covariance matrix* option is still selected. Deselect the *Output: e values* option. Select the *Output:X matrix* option.

5. Click the *Compute* button. In a few seconds, a matrix form containing the results matrix will appear on the desktop. This matrix should be titled *x_matrix.dat*.

6. Create the following two graphs:

    (1) The graph of column 1 of matrix *sinnois1.dat*, the original data matrix: Click the matrix button corresponding to this matrix (it should be button 1). From the *Graph* menu, select the *XY Graph* option. On the *XY Graph* dialog form, enter 1 in the *Y axis: End column* field. Click the *New Graph* button. A new graph will be created that should resemble Figure 3.8, showing one cycle of a sine wave with added pseudo-random noise.

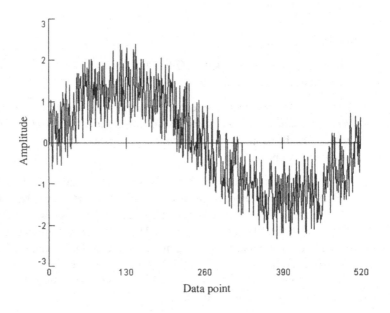

Figure 3.8. The graph shows column 1 from matrix *sinnois1.dat*, consisting of a 1-Hz sine wave with added pseudo-random noise.

    (2) The graph of column 1 of matrix *x_matrix.dat*, using the following procedure: Click the matrix button corresponding to this matrix (it should be button 2). On the *XY Graph* dialog form, enter 1 in the *Y axis: End column* field. Click the *Redraw* button. The graph will be redrawn and the result should resemble Figure 3.9.

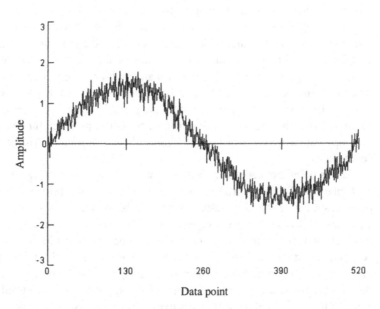

Figure 3.9. The graph shows column 1 from matrix *x_matrix.dat*, the result of the principal component filtering operation, showing the one-cycle sine wave with clearly less noise than was present in the original data.

The new graph shows one cycle of a sine wave with a clearly higher signal-to-noise ratio than that shown in the first graph. The principal component filtering has accomplished its objective of removing a significant amount of noise from the original data.

Principal Component Data Reduction

Introduction

A second application of the principal component function is to deal with correlated variables. Correlated variables can be a handicap when they are included in the set of independent variables used to train a neural network. For example, if there are two correlated variables in the network training data, the second variable gives the network no new information. On the other hand, the second variable increases the amount of data that the neural network must deal with in trying to model the functional relationships within the data.

One way in which correlated variables can be identified, and dealt with, is to carry out a principal component analysis on the predictor portion of the network training matrix. Recall that a network training matrix consists of a number of rows, each row representing a single exemplar, and a number of columns. If each exemplar consists of the values of n independent and m dependent variables, the predictor portion of the matrix is contained in the first n columns of the training matrix. The result of carrying out the principal component analysis will be a matrix with a reduced number of columns. Significantly, these columns will be mutually uncorrelated. To this matrix, representing the new predictor portion of the network training matrix, are then simply added the criterion columns from the original training matrix. Simulnet provides functions to add columns and copy data across sets of matrix columns. These operations will be described in more detail below.

At this point the reader may want to review the section dealing with principal component analysis. To recap briefly: given a data matrix, the principal component analysis generates a list of the eigenvalues of the matrix. A matrix of n columns, representing n variables, will have n distinct eigenvalues. The size of each eigenvalue represents how much of the variance in the original matrix can be accounted for by means of some linear combination of the variables, or columns, in that matrix. If all of the variables (columns) in the matrix are mutually uncorrelated, the n eigenvalues will be of roughly the same order of size. If, on the other hand, some of the variables are correlated with others, then some of the n eigenvalues will be relatively very small. Suppose a data matrix consists of three columns, containing values sampled from three corresponding variables. If the three variables are mutually uncorrelated, the three eigenvalues of this matrix will be of the same order of size. However, if the data in one of the columns, say column 2, is correlated with the data in, say, column 1, then two of the eigenvalues of this matrix will be of approximately equal size. The third eigenvalue will be relatively much smaller, signaling the fact that two of the matrix columns were correlated.

Using principal component analysis is, as described in the previous section, a two-stage operation. In the first stage, all of the eigenvalues for the predictor portion of the network training matrix are computed, and a decision is made about how many of these eigenvalues to retain. The total number of eigenvalues provided by the principal component analysis is equal to the number of predictor columns in the network training matrix. The principal component analysis will present the eigenvalues sorted in order of decreasing size. Out of this total number, a subset of the largest eigenvalues is selected. If the network training matrix contains 24 predictor columns, for example, the analysis will provide 24 eigenvalues. It might be decided that only the 12 largest of these eigenvalues are significant and will be kept. In some cases, it may not be easy to decide how many eigenvalues to keep. As well, the principal component analysis may take a significant amount of time to carry out if there is a large number of predictor columns. In sum, by inspecting the sizes of the eigenvalues and applying the heuristics described earlier, a decision is made about how many eigenvalues to retain.

In the second stage, the principal component analysis is asked to generate a matrix containing a number of columns equal to the number of retained eigenvalues. This matrix is referred to as the standardized component scores matrix, symbolized by $\mathbf{Z}$. The columns of matrix **Z** represent new variables that are mutually uncorrelated, each of which is a linear combination of the original predictor variables. If matrix **Z** is used as the predictor portion of the network training matrix, the network now has fewer data points to deal with, data points each of which now carries independent information about the functional relationships within the original training matrix.

Simulnet Exercise: Principal Component Data Reduction

In this exercise a principal component data reduction operation will be performed on the matrix contained in file *phase.trn*. This matrix will be referred to as the data matrix, in order to generate a matrix of standardized principal component scores **Z**. If the data in the columns in the data matrix are to some extent mutually correlated, it can be expected that some of the eigenvalues of this matrix will be relatively small. By retaining only the largest eigenvalues, matrix **Z** will contain a smaller number of columns than the data matrix. This reduced number of columns represents a set of mutually uncorrelated composite variables created by taking linear combinations of the original variables.

Procedure

1. Close all forms on the desktop. Open file *phase.trn* and minimize the matrix form. This matrix will be referred to as the data matrix.

2. From the *Analyze* menu, select the *Principal Component Analysis* option. On the *Principal Component Analysis* dialog form, enter a value of 24 in the *Data matrix: Ending column* data entry field. Enter a value of 24 in the *Computations: No. of retained components* field.

3. Click the *Compute* button. After a few seconds, a matrix form containing the results matrix will appear, titled *e_values.dat*. The first and last three values in this matrix should be close to those shown in the following table (rounded to three significant figures). Each row contains the magnitude of one of the eigenvalues of the data matrix. Note that only the first eigenvalue is greater than one. For the purpose of this exercise it will be assumed, arbitrarily, that the first 12 eigenvalues are significant. After inspecting matrix *e_values.dat*, close this matrix form to remove it from the desktop.

Eigenvalues
7.07
0.606
0.19
:
0.0114
0.00773
0.00678

4. Click matrix button 1 to ensure that the data matrix has the focus. Deselect the *Output: e-values matrix* option. Select the *Output: Z matrix* option. Enter a value of 24 in the *Data matrix: Ending column* data entry field. Enter a value of 12 in the *Computations: No. of retained components* field.

5. Click the *Compute* button. In a few seconds, a new matrix, *z_matrix.dat*, will appear. Close the *Principal Component Analysis* dialog form. The desktop should now contain only two matrices: *Matrix 1: phase.trn*, and *Matrix 3: z_matrix.dat*.

   Matrix 3: z_matrix.dat contains 12 columns. Each of these columns represents a new variable, constructed by taking a linear combination of the variables in the original data matrix. Furthermore, each of these variables has been constructed in such a way that they are all mutually uncorrelated. By using these new variables as the predictor variables in a network training matrix, it can be confidently expected that the network is training on nonredundant data.

6. To finish constructing the new network training matrix, the criterion values from the original network training matrix, *Matrix 1: phase.dat* must be added. Click matrix button 1 to give this matrix the desktop focus and restore the matrix to normal size so that the contents can be viewed. Scroll over to the last column, column 25, of the matrix. Select the entire column by clicking on the column heading (the cell at the top of the column containing the numeral 25). Then, copy this column by pressing CTRL-C, or by selecting *Copy* from the *Edit* menu. The *Matrix Copy* dialog form will appear. Click the *Copy to Copy Buffer* button on this form. This will copy the contents of column 25 to the Simulnet copy buffer. Minimize matrix 1.

7. Next, click on matrix button 3 to shift the desktop focus to *Matrix 3: z_matrix.dat*. From the *Transform* menu select the *Matrix Tools* option. On the *Matrix Tools* dialog form select the *Add column* operation from the *Operations* list. Enter a value of 13 in the *Parameters: column number* field. Click the *Compute* button. A new matrix will appear on the desktop, titled *z_matr_1.dat*. This matrix will be identical to *Matrix 3: z_matrix.dat*, but with an extra column. Close the *Matrix Tools* dialog form. Close matrix 3, and answer NO to the question box that will appear.

8. Restore *Matrix 4: z_matr_1.dat* to normal size so that the contents of the matrix form can be seen. Scroll to column 13. Select the entire column by clicking on the column header (the cell at the top of the column with the numeral 13). Select the *Paste* option from the *Edit* menu. On the *Matrix Paste* dialog form, select *Paste from Copy Buffer*. The contents of the Simulnet copy buffer will now be pasted into column 13. Inspecting this column, it should be identical to the criterion column, column 25, in the original data matrix, *matrix 1: phase.trn*.

9. If desired, matrix 4 can be saved, using a *trn* extension to identify it as a network training matrix. After it has been saved, this matrix can then be used to train any of the networks. For example, it can be used to train the *Probabilistic Network*, using the *jackknife* option.

4
A Data Analysis Protocol

Introduction

This section presents a protocol, a systematic procedure, that can be used as a guide through the steps involved in preprocessing and analyzing experimental data. In contrast with most other sections of this book, the present section is designed be applied to the reader's own data.

This section is divided into two parts. The first part provides a checklist outlining a series of data preprocessing options. The second part deals with the process of actually analyzing the data once any preprocessing steps have been completed. In the work of this section the reader has the option of using files that have been provided, or using their own data.

A Preprocessing Checklist

Data preprocessing includes operations that are performed on experimental data in order to enhance or select a sub-set of the data features or to deal with artifacts presumed to be present in the data. Data might be preprocessed, for example, in order to select a particular frequency band, or to remove features that are artifacts of the collection process rather than information related to the system being measured. This procedure is organized as a series of steps, each of which asks a question. The question may involve the data itself, the experimental paradigm within which the data was recorded, or any hypotheses the reader may have about the data. These questions can also be applied to observational or historical data, as well as to experimental data.

1. Is the experimental manipulation expected to have an effect on the mean value of the data recorded from each system variable? Unless there are principled reasons for believing otherwise, the answer here should be *no*. If the answer is no, set the means from all columns of all data matrices to zero using the *Detrend - Order 0* function.
2. Is the experiment expected to have an effect on the variance (very roughly, variations in amplitude) of the data? That is, Is the experimental manipulation expected to result in differences in the amplitude of the data between different conditions?

Again, unless there are reasons for believing that this will be the case, the answer is *no*. If the answer is no, standardize all of the data matrices. Standardizing will eliminate differences between the data files in terms of variance, and in terms of mean (there are procedures for removing only differences due to variance and not those due to mean, but in this procedure it will be assumed that if differences in variance are not relevant, neither are differences due to mean). Use the *Standardize Columns* function to standardize the data.

3. Is the experimental manipulation expected to have most effect on one frequency band? If there are reasons to believe that it will, filter the data to extract that band of frequencies using one of the Simulnet *Filter* functions. If in doubt, or if there is no reason to believe that the experiment will affect one band of frequencies more than another, then do not filter the data. Many sensitive analytical procedures such as neural networks have most chance of finding a between-condition difference if they are presented with all of the frequency components.

4. Is the data expected to contain outliers? Outliers are data values that exceed the mean value of the data by some large factor. This factor is generally expressed in terms of *standard deviations*. For instance, an outlier can be defined to be any data point that has a value greater than ± 4 standard deviations from the mean. Generally, if the experimental procedure was not expected to produce extreme values, values greater than, say, 4 standard deviations from the mean, then if such outliers do nevertheless exist, they should be eliminated from the data. While neural networks can deal with extreme data values, such values can slow the rate at which a network is able to learn to model the data. When a significant number of outliers are present, a network may be effectively prevented from learning at all. All of the network functions include a function that can be invoked to carry out an analysis on the training data. This analysis will, among other things, examine the data for outliers, and will offer to trim those outliers.

To trim outliers from data, open the file containing data. Then, select the *Transformations* option from the *Transform* menu. On the *Transformations* dialog form, select the *Operation: Trim Outliers* function. Enter a value of 4 in the *Parameters: number of std. dev.'s* field. Click the *Compute* button. A new matrix form will appear, containing a copy of the original data with outliers trimmed. This new matrix should be saved using some name that distinguishes it from the original one.

Analyzing Experimental Data

This section will present a sample data analysis, using a set of three exercises. The exercises will assume that the data contains two categories, and that the goal of the analysis is to see if a neural network can classify the data into those categories. In the first of the following exercises, the data will be inspected by graphing it. In the second exercise, as a preliminary analysis, the data will be examined to see what frequency components are present. In the third exercise, the data will be used to create a network training matrix. A neural network will then be trained to classify the data.

Simulnet Exercise: Inspecting the Data

It is good practice to confirm that the data matrix to be analyzed is organized in the proper format. Simulnet functions expect that matrix columns represent variables, while rows represent cases or data points. Graphing the data is one way to confirm that this is so (assuming the file owner knows what the file contents should look like). Another very important reason for graphing the data is to confirm that it doesn't contain features that are known to be artifacts.

This exercise can be carried out using either the reader's own data file, or the pre-prepared data file *eeg00.dat*. This file contains one second of EEG data recorded from 16 channels, at a sampling rate of 128 samples per second. If a different data file is used, the file format must be either in text-only (ASCII), or Simulnet binary format, and the file must organized as a matrix with columns representing variables and rows representing cases.

Procedure

1. Close all forms on the desktop. Open the data file as a matrix and minimize the matrix form.
2. Select the *XY-Graph* option from the *Graph* menu, or click the XY Graph button on the toolbar. On the *XY-Graph* dialog form, select the *Attributes: color* option. Enter a value of 1 in the *Data Matrix: End column* data entry field.
3. Click the *New Graph* button. A graph form will appear containing the graph of column 1. If file *eeg00.dat* is being used the graph should look like Figure 4.1.

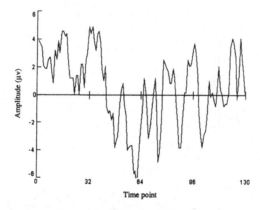

Figure 4.1.The graph shows the matrix in file *eeg00.dat*, a 1 second recording of EEG data from 16 channels. The y-axis represents voltage level in microvolts, and the x-axis represents time-point in 1/128th's of a second.

4. After examining the graph, close the graph without saving it. Close the *XY-Graph* dialog form as well. The desktop should only contain the matrix form containing the data matrix.

Simulnet Exercise: Fourier Analyzing the Data

This exercise will assume that the general question being asked is the following: What frequency components does the data contain? In other words, what does the frequency spectrum of the data look like? The analytic procedure that is appropriate to this question is Fourier analysis. In this exercise the first column of the data matrix will be Fourier analyzed, generating a frequency spectrum of the data in that column.

Procedure

1. Select the *Fourier transform* option from the *Analyze* menu. On the *FFT* dialog form, enter a value of 128 in the Parameter: Sample rate field. Enter a value of 1 in the *Data Matrix: End column* data entry field. Click the *Compute* button. A matrix form containing the frequency spectrum will appear. Column 1 of this new matrix contains the actual frequency spectrum, that is, the power level at each frequency. Column 2 of this matrix contains the corresponding frequency values.
2. Close the FFT dialog form, and close the data matrix. The desktop should now only contain the matrix form containing the frequency spectrum.
3. Graph the spectrum using the following procedure: Click the matrix button corresponding to this matrix. On the *XY-Graph* dialog form, enter 1 in the *Y-axis: End column* field. De-select the *X-axis: Use row no.* option, and enter 2 in the *X-axis: Use column* field (the reason for these settings is that the spectrum matrix uses column 1 to hold the frequency spectra, and column 2 to hold the values of frequency). Click the *New Graph* button. A new graph should be created. If file *eeg00.dat* is being used for this exercise, the graph should resemble Figure 4.2.

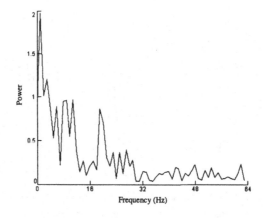

Figure 4.2. The graph shows frequency spectra for column 1 of matrix *eeg00.dat*. EEG data generally has a spectrum that looks like this: the greatest amount of power occurs in the frequency range of approximately 0 to 16 Hz.

Simulnet Exercise: Neural Network Classification Analysis

In this exercise the general question being asked is the following: Is there a between-condition difference in the data? In other words, are the effects of the experimental manipulation reflected in differences in the data recorded in the different conditions? One analytic approach to dealing with a question of this sort is neural network classification. This approach addresses the more specific question, Can a neural network be trained to distinguish between data that were collected in the different conditions in the experiment?

This exercise will use a set of 20 data files recorded in an experiment designed to investigate human visual perception. Each file contains a 1 second recording of 16 channels of EEG data. Each of these 16 channels was recorded from 1 of 16 scalp electrodes. The data was recorded with a sampling rate of 128 samples per second. This gives, in each of the 20 files, a matrix of 128 rows, one for each time-point, by 16 columns, one for each EEG channel. These 20 files were recorded in an experiment that involved 2 conditions. Files *eeg01.dat* through *eeg10.dat* were recorded in condition 1, and files *eeg11.dat* through *eeg20.dat* were recorded in condition 2. In condition 1 subjects were asked to look at an image containing a camouflaged object, while in condition 2 subjects were asked to look at a blank screen. Each of these data files constitutes a single network training exemplar. A network training file containing these exemplar files has been provided, as file *eeg.trn*.

Creating a Neural Network Training File

Before the experimental data can be used to train a neural network, the individual data files must be combined into a single network training file. A network training file is a file that is made up of a set of exemplars which presumably contain the relationship that the network is going to be asked to learn. Each exemplar comes from one of the data files generated by the experiment.

Structurally, each row of the network training matrix contains one network exemplar. Each exemplar consists of two sections. The first section contains the values of the independent variables, the data from one of the data files that are to be used. Each of these independent variables is in fact one of the data points from one of the data files. Each of these files consists of 128 time-points by 16 channels, giving a total of 2048 data points. This results in 2048 independent variables in the first section of an exemplar. The second section of each exemplar contains the value of a single, dependent, variable. In the present exercise, this dependent variable is simply a dummy code, either a 0 or a 1, encoding the experimental condition to which the exemplar belongs. If an exemplar corresponds to condition 1, the dummy code is 0, otherwise the dummy code is 1. The entire exemplar then consists of 2049 values, 2048 of which are the values of independent variables, and 1 of which is the value of a dependent variable, the dummy code. The network will be asked to learn the relationship between these independent and dependent variables.

These 20 exemplar files next need to be combined into a network training file. This is done using the procedure that was presented previously for creating a training file. This procedure is presented here in abbreviated form:

1. Select the *Create Training File* option from the *Network* menu. On the *Create Training File* dialog form click the *Example File: Select* button. This will show the *Select Files and Paths* dialog form. Select the 20 exemplar files *eeg01.dat* through *eeg20.dat*.
2. Select the path into which the training file is to be placed
3. Enter a value of 1 in the *Criterion: No. of columns* field.
4. Click the *Next* button. The *Criterion Values* form will appear on the desktop, showing a matrix with one column for each criterion variable, and one row for each example file. For each example file, enter a dummy code to label the exemplars as to whether they belong to condition 1 or 2. For files *eeg01.dat* through *eeg10.dat*, enter a dummy code of 0. For files *eeg11.dat* through *eeg20.dat*, enter a dummy code of 1.
5. When the dummy codes have been entered, click the *OK* button on the *Criterion Values* dialog form. The training matrix will now be constructed, a process that may take several seconds to complete. When the matrix has been created, a *Save File* dialog form will appear. Save the training file as file *eega.trn*, in the directory in which Simulnet was installed. The training file is now ready to be used.

Using the Probabilistic Network

1. From the *Network* menu, select the *Probabilistic Network* option. On the *Probabilistic Network* dialog form make the following changes: For *Examples: Criterion columns* enter a value of 1. Click the *File: Training data* button. This will show the *Load Train File* dialog form. Select the file *eeg.trn*, and click the *OK* button. It will take a few moments to load the file and perform various updates on the *Probabilistic Network* dialog form. Select the *Testing data: Jackknife test mode* option. For *Testing Parameters: Smoothing* enter a value of 4 (this value may already be set).
2. Click the *Compute* button. The network may take from several seconds to several minutes to complete its operations, depending on the size of the training file. When the network has finished its computations, a matrix form will appear containing the results matrix. If desired, this results matrix can be saved to disk. The values in the results matrix should resemble those in the following table (rounded to 3 significant figures). Each row corresponds to one matrix exemplar, and thus to one of the original data files. Column 1 is the dummy code assigned to the exemplar to mark the condition that it represents. The first 10 exemplars correspond to files from condition 1 (dummy-coded as 0). The next 10 exemplars correspond to files from condition 1 (dummy-coded as 1).

Network Scores

Dummy code	Score
0	0.725
0	0.601
0	0.588
0	0.533
0	0.334
0	0.342
0	0.677
0	0.311
0	0.538
0	0.346
1	0.972
1	0.819
1	0.740
1	0.775
1	0.589
1	0.417
1	0.497
1	0.708
1	0.550
1	0.278

3. Close the *Probabilistic Network* dialog form. The Simulnet desk-top should now only contain the network results matrix.

Assessing Network Results

The network results matrix will have two columns, and a number of rows equal to the number of original data matrices that were selected for inclusion in the training matrix, in this case 20. Recall that each of those original data matrices was used as one network training exemplar. Since the Jackknife test mode was used, each of the exemplars was tested against all the other exemplars. In the Jackknife test mode each exemplar is tested by first removing the exemplar from the training matrix. The network is then trained on the remaining exemplars. After the network has finished training, the exemplar that was removed is presented to the network for testing, and the network is asked to classify that exemplar. This procedure is repeated for each exemplar in the training matrix.

The results of this operation are contained in the network results matrix. Each row of this matrix represents one exemplar in the training file. Column 1 shows the dummy code that was assigned to each exemplar in the training file when the file was first created. Some exemplars are coded with a 0, indicating that they belong to the first experimental condition. The remaining exemplars are coded with a 1, indicating that they belong to the second experimental condition. These dummy codes are *only* shown in this network results matrix for the user's convenience. The network does not use these codes in any way. Column 2 of the network results matrix shows the actual score assigned by the network to each of the exemplars.

In order to quantify how well the network was able to classify the training file exemplars, a statistical test will be carried out on the scores. One straightforward test that can be applied is the Student's t-test. If the value of probability associated with the test is less than some critical value, such as 0.05, it could be concluded that the network would adequately classify the exemplars in the training file. The corresponding substantive conclusion would then be that the experimental manipulation has had a significant effect on data recorded within the different conditions.

Applying the Student's t-test

This procedure assumes that the scores dummy-coded as 0, and the scores dummy-coded as 1 are located in contiguous rows of the network results matrix. If this is not the case, the matrix rows will need to be rearranged by creating a new matrix (see procedure given earlier), and copy one row at a time from the network results matrix, to the new matrix, so that all of the condition '0' rows are in one group in the top rows of the matrix, and all the condition '1' rows follow.

1. From the *Analyze* menu, select the *Statistics - Inferential* option. On the *Inferential Statistics* dialog form, click on the *Type of analysis* list-box to show the list of available options, and scroll through the list to find the *t-test (unequal variance)* function. Select the *Output: text* option. For *Data group 1: starting column* enter a value of 2. For *Data group 1: ending column* enter a value of 2. For *Data group 2: starting column* enter a value of 2. For *Data group 2: ending column* enter a value of 2. For *Data group 1: starting row* enter a value of 1. For *Data group 1: ending row* enter a value of 10. For *Data group 2: starting row* enter a value of 11. For *Data group 2: ending row* enter a value of 20.

2. Click the *Compute* button. A text-box containing the results of the analysis will appear. This text-box should contain information similar to the following:

T-Test of Distributions with Equal Variances

Item	Group 1	Group 2
Starting row	1	11
Ending row	10	20
Starting column	2	2
Ending column	2	2
Number of cases	10	10
Mean	0.499	0.634
Variance	0.024	0.043
df	18	
Pooled variance	0.033	
t-value	-1.652	
probability:		
one-tailed	0.058	
two-tailed	0.116	

3. The critical value that should be examined is *probability: one-tailed*. If this value is less than or equal to approximately 0.05, the difference between the means of the two groups could have arisen by chance less than 1 time in 20. This criterion is generally accepted as being the minimum requirement for considering that the two groups are significantly different. If the one-tailed probability is less than 0.05, it may be concluded that the experimental manipulation appears to have resulted in a significant effect in the recorded EEG. If the value is between 0.1 and 0.05, the results may be referred to as being marginally significant. If the value is greater than 0.1, then one of two things may be true. Either the effects of the experimental manipulation on the EEG are not significant, or there are effects, but the network classification analysis has not been powerful enough to detect them. As in all statistical testing, absence of evidence can never be taken to be evidence of absence.

    In the present case, the value of one-tailed probability is 0.058, a marginally significant outcome. It could be concluded that there appears to be a marginally significant difference between the two groups of network scores: those for exemplars from condition 1 and those for exemplars from condition 2. There is some evidence that the network has been able to learn to distinguish the samples of EEG data that correspond to condition 1, viewing a camouflaged object, from those corresponding to condition 2, viewing a blank screen. In other words, there does appear to be a difference in the EEG data corresponding to these two conditions. This tentatively positive result should motivate more detailed questions of the data, related to which features in the data are associated with the difference between conditions.

4. When you have finished with the *Inferential Statistics* dialog form, click the *Close* button to remove the form from the desktop. When you have finished with the textbox, close this form as well.

Glossary

attractor An attractor is a description of the behavior of a dynamical system. The attractor exists in a space of dimensionality equal to the number of defining variables of the system. A one-dimensional time series can be used to construct an artificial n-dimensional attractor by plotting n-tuples of data points separated by some fixed number of points (the lag). If the time series is an adequate sample of the activity of some n-variable process, then the dimensionality of the artificial attractor is an estimate of n.

backpropagation More accurately termed error backpropagation (EBP), this is an algorithm that is commonly used to train neural networks. Network weights are modified in a way that will minimize the sum of squared errors computed over all output nodes. The EBP algorithm is a form of gradient descent: The network descends, as it were, within the landscape defined by the network weights where height corresponds to squared error, to a point of minimum error. The probability that this point is only a local minimum will decrease as the dimensionality of the weight space — as the number of weights — increases.

coherence Coherence is an index of the degree of similarity in the patterns of the frequency components of two data sets. Two data sets whose frequency components are aligned, or 'in step', will have a large value of coherence.

conformable Two matrices are conformable if the number of columns in the first matrix is equal to the number of rows in the second matrix. Two matrices can only be multiplied together if they are conformable.

Example: A matrix **A** with 3 rows and 5 columns and a matrix **B** that has 5 rows and 2 columns are conformable and can be multiplied, in that order, as **A** x **B**. Note that if the matrices are considered in the opposite order, they are not conformable, and can not be multiplied. In this example, the operation **B** x **A** is not defined, and can not be carried out.

correlation There are many indices of correlation, all specifying the degree of association between two data sets. When not further qualified, this term generally refers to the Pearson product moment correlation, a measure of the degree of linear association between two data sets.

The magnitude of correlation depends only on how closely matched are the patterns of behavior represented by the two data sets. Magnitude of correlation does not depend on the relative amplitudes of the patterns.

correlation dimension

Correlation dimension is one measure of the fractional dimension of a phase space attractor. The correlation dimension is an estimate of the lower bound on the number of variables of the dynamical system that generated the time series from which the attractor is constructed.

correlation integral

The correlation integral describes the probability of finding a pair of points on an attractor within some distance of each other, averaged over all or a sample of the points on the attractor. The slope of the graph of the log of the correlation integral vs. the log of the distance is the correlation dimension for the attractor.

covariance

Covariance is an index of the extent to which two data sets covary; that is, how closely matched, data point by data point, are the behaviors represented by the two data sets. The magnitude of covariance depends not only on the pattern of behaviors but also on their amplitude.

diagonal elements

For a square matrix A, the diagonal elements are the elements a_{ii}, where i refers to both rows and columns.

diagonal matrix

A diagonal matrix is a square matrix with non-zero diagonal elements, and with all other elements zero.

eigenvalue

This term is applied to a matrix of p columns, each column of which represents a single measured variable. A matrix of p mutually independent columns will have p distinct eigenvalues. If, for example, the matrix has 3 columns the data in the matrix can be visualized as a set of points in a 3-dimensional space. This space will be referred to as the embedding space. The shape defined by this set of points will in a simple case be an ellipsoid. The 3 mutually perpendicular axes of this ellipsoid will in general not be parallel to the axes of the embedding space. The lengths of the three axes are specified by the magnitudes of the three eigenvalues of the matrix. The largest eigenvalue specifies the length of the major axis of the ellipsoid. The second and third largest eigenvalues specify the lengths of the second and third axes of the ellipsoid.

eigenvector

(See first description of eigenvalue) For each unique eigenvalue of a matrix there is a corresponding eigenvector. Graphically, each eigenvector specifies the angle (actually the cosine of the angle) between each of the axes of the embedding space and the axes of the ellipsoid. Assume again a 3 dimensional embedding space. The eigenvector corresponding to the largest eigenvalue specifies the (cosines of the) angles between the major axis of the ellipsoid and the 3 axes of the embedding space. The remaining 2 eigenvectors correspondingly specify the cosines of the angles between the two minor axes of the ellipsoid and the 3 axes of the embedding space.

entropy

Entropy is a measure of the average information content of a set of data points, computed as the sum of the information content of each data point $I(x)$ weighted by its probability $P(x)$. For a set of points $X = (x_1, ..., x_i, ...)$, entropy is $H(X) = \Sigma_i P(x_i) I(x_i)$. Also referred to as Shannon entropy.

genetic
algorithm

Genetic algorithms are a general approach to locating a global minimum (in terms of some error function) in the parameter space of a problem or task. A set of parameter values is treated as an artificial analog of a chromosome. A population of such chromosomes is initially created using random values. Each chromosome is tested for fitness using a measure of fitness, a fitness function, appropriate the problem at hand. The fitness function evaluates how well a chromosome is able to solve the problem or carry out the task. Individual chromosomes are then allowed to pair and mate by exchanging portions of the information they carry (crossover), with a probability proportional to the fitness of each chromosome. The result is a new population that replaces the parent population. Elements of each chromosome can also be randomly changed (mutated) with a small defined probability. The new population is now evaluated for fitness. The process of artificial evolution is repeated until a chromosome appears that meets some criterion for fitness.

generalized
regression
neural network
(GRNN)

Like the BackProp network, the GRNN is a function-approximation algorithm. The GRNN is trained by presenting it with a set of training vectors representing an instance of the function to be learned or approximated. The GRNN is structurally and functionally related to the PNN. It shares with the PNN the advantages of a fast, single-pass training phase, and relative insensitivity to outlying data values in the training set.

information

The information content of a data point is inversely proportional to its probability. Thus, the less probable — the more surprising — the data point, the higher its information content. A completely predictable datum, with a probability of 1, has an information content of zero. The datum tells us nothing new. More precisely, the information content of a datum x is the logarithm of the inverse of the probability of the datum: $I(x) = \log (1/P(x)) = -\log P(x)$. If the logarithm is taken to base 2 then the units of information are bits.

The information content of a set of data points is specified by its *entropy*.

Learning vector quantizer (LVQ)

The Learning Vector Quantizer is supervised classification algorithm, the LVQ as a supervised version of the SOM. Rather than allowing the network nodes to self-organize, the network is guided to develop reference vectors representing the training categories. For each training vector the most similar reference vector is identified. The reference vector is then modified according to whether or not it belongs to the same category as the training vector. Reference vectors are pulled towards training vectors of like categories and pushed away from training vectors of unlike categories. After being trained, the LVQ can be used to classify novel vectors.

matrix

A matrix is set of numbers, or elements, organized into rows and columns. Matrix columns are assumed to represent variables, and matrix rows to represent the individual cases corresponding to each of those variables. Matrices are denoted by upper-case boldface letters, for example, $\mathbf{X}$.

Each place in the matrix at the intersection of a row and column, where a single value is stored, is referred to as a cell. The value in matrix $\mathbf{X}$ at the intersection of the i-th row and the j-th column is denoted by x_{ij}.

mutual information

Applied to two sets of data points, mutual information is a measure of how much information can be gotten about one of the sets by examining the other set. Mutual information is an index of the degree of general association between the two data sets.

neural network

A neural network is a device, generally implemented in software, that can learn (approximate) a functional relationship between a set of predictor (independent) and criterion (dependent) variables. Information about the relationship is distributed among a set of exemplars each of which instantiates a single example of the relationship. After being trained on the exemplars the network can supply the value of the dependent variables, given the independent variables. Structurally, neural networks consist of nodes connected by trainable links or weights. Input nodes receive the values of the independent variables, and are connected through a set of weights to one or more sets of hidden nodes. Hidden nodes learn the features in the training data, and are connected through a second set of weights to output nodes that provide the network's estimates of the values of the dependent variables. Over the course of training, the values of the network weights are modified using an algorithm such as error backpropagation.

node

In the context of a neural network, a node is a simplified model of a neuron. Nodes receive a number of input signals x_i, and generate one output signal y. The output signal is a nonlinear function of the summed inputs: $y = f(\Sigma x_i)$, where f(.) is some nonlinear function such as the sigmoidal or hyperbolic tangent function.

order

The order of a matrix is the size of the matrix in rows and columns. The order of a matrix with 3 rows and 4 columns is 3 by 4.

orthogonal

Perpendicular; orthogonal line segments lie at right angles to each other: $\perp$. When applied to 2 or more matrix columns, it means that the data within these columns is mutually independent, or uncorrelated. The behavior of the variables represented by these columns is thus uncorrelated.

phase

For a single cycle of a periodic waveform, phase is a measure of distance along the cycle. Phase can be expressed using a number of different units including degrees, radians, and grads. In degrees, a single cycle extends from a phase of 0 to 360 degrees. Thus, the instantaneous amplitude a of a sine wave of frequency f and maximum amplitude A is given by the relation

$a = A \sin (2\pi f + \varphi)$, where φ is the phase angle.

For two or more periodic waveforms of the same frequency, the relative phase, or phase difference, is a measure of how well the features of the waveforms coincide. A phase difference of 0 degrees indicates that the two waveforms are perfectly lined up. A phase difference of 90 degrees indicates that the two waveforms are offset relative to each other by a quarter cycle.

probabilistic
neural network
(PNN)

A probabilistic neural network is a classification algorithm. Given a set of training vectors belonging to one of some number of categories, and whose category membership is known, a PNN will assign novel (test) vectors into the most probable of the categories. An advantage over a backpropagation network is speed: There is no equivalent to the backpropagation network's training phase. Testing novel vectors can however take a significant amount of time with a large number of training vectors. PNN's are generally insensitive to outlying data values in the training set.

singular

Applied to a matrix, the term singular means that the rows or columns of the matrix are not independent. This would be the case if one column of a matrix was a linear combination of other columns of the matrix. As an example, if one column of a matrix was equal to another column of that matrix multiplied by some factor, the matrix would be singular.

Self-organizing
map (SOM)

A self-organizing map is an unsupervised vector classification algorithm. A network of nodes, each representing a reference vector, is trained by presenting it with a set of training vectors. The network self-organizes so that clusters of nodes come to represent the categories within the training set. Following training the SOM can be used to classify novel vectors.

square matrix

A square matrix is a matrix that contains an equal number of rows and columns.

symmetric
matrix

A square matrix with equal values for corresponding elements above and below the diagonal. For a square matrix A, the values of elements a_{ij} will equal the values of elements a_{ji}. A correlation matrix is an example of a symmetric matrix.

time series

A time series is set of data points collected over some period of time. Generally, such data points are collected at regular time intervals. The rate at which the data points are collected is referred to as the sampling rate.

transpose

To transpose a matrix is to interchange the rows and columns. For example, transposing the matrix Y

$$\begin{bmatrix} 2 & 6 & 1 \\ 5 & 3 & 7 \end{bmatrix}$$

produces the matrix Y':

$$\begin{bmatrix} 2 & 5 \\ 6 & 3 \\ 1 & 7 \end{bmatrix}$$

For matrices containing data in which the variables are contained in rows rather than in columns, Simulnet provides a transpose operation. Thus, a matrix organized with variables in rows can be first transposed so that variables are contained in columns, operated on by Simulnet functions that expect variables to be in columns, and then transposed once again to return the matrix to its original format.

vector

A vector is a set of numbers, or components, that together describe a variable or a case. One matrix row can be considered to be a vector, a row vector, describing one case. One matrix column can also be considered to be a vector, a column vector, describing one variable. A vector is denoted by a lower-case boldface character, e.g., x. As an example, a person could be described in terms of a height-weight vector. Such a vector would have two components, the person's height and the person's weight. For example, the vector $x = (165, 55)$ describes a person whose height is 165 cm and weight is 55 kg.

Index

Activation function, 23, 24, 50
Albano, 135, 139, 141
Algorithm
  backpropagation, 1, 26, 28, 29, 30,
    31, 32, 53, 75, 76
  QL, 191
  supervised classification, 216
  unsupervised classification, 217
Aliasing, 174
Alkon, 75
Almeida, 75
Analysis
  coherence and phase, v, 2, 153, 154
  correlation, 14, 160
  correlation dimension, 128
  covariance, v, 2, 14, 163
  eigenvalue, v, 2, 14, 146, 147, 149,
    152
  Fourier, v, 2, 14, 15, 24, 51, 144,
    206
  fractal dimension, v, 2, 14, 128
  mutual information, v, 2, 14, 160,
    162
  neural network, 2, 75, 164, 181, 182
  principal component, 181, 189, 191,
    192, 193, 195, 196, 200
Angus, 38, 73
Antonsen, 141
Attractor, 71, 126, 130, 132, 133, 134,
    135, 136, 137, 139, 213, 214
Autocorrelation, 135

Babloyantz, 140
Backpropagation
  algorithm, 1, 26, 28, 29, 30, 31, 32,
    53, 75, 76
  algorithm, error surface, 26

computing weight changes, 29
learning rule, 13, 25, 59, 67, 87
training, 38, 86
training mode, 83, 84, 86, 88, 89
Badii, 137, 139
Baker, 70, 73
Baum, 40, 73
Beauregard, 153
Berliner, 130, 139
Bioch, 51, 73
Blume, 132, 140
Bondeson, 137, 141
Broggi, 137, 139
Broomhead, 138, 139
Buzaki, 55, 58, 74

Casdagli, 139, 140
Chaos, 70, 130, 136, 138, 139, 140,
    141
Chaotic
  attractor, 136, 140
  behavior, 30, 129, 130, 137
  component, 130
  dynamics, 130
  process, 130
  processe, 130
  system, 70, 129, 130, 131
  time series, 67, 69, 100, 101, 111,
    112, 123, 124, 125, 126, 127,
    128, 141
Chatterjee, 130, 139, 140
Chester, 41
Ciliberto, 137, 139
Classification
  accuracy, 57
  algorithm, 216, 217
  Bayesian, 94

EEG, 57, 58, 73
function, 98
jackknifed, 43
LVQ, 58
mis-, 56
neural network, 113, 114, 121, 181, 207, 212
of data vectors, 79
of exemplars, 24
pattern, 24, 39
performance, 55, 114
problem, 17, 121, 122
procedure, 80
rate, 111
rule, 80
score, 80, 114
task, 13, 17, 24, 25, 46, 119, 120, 121, 126
techniques, 51
Classifier, 75, 91, 94, 97, 98
pattern, 1, 58
Coherence, 2, 14, 51, 153, 154, 155, 156, 157, 158, 159, 168, 213
Complexity, 2, 14, 50, 128, 129, 138, 142, 144, 146
Computing power, 56, 172, 182
Confidence interval, 154, 155
Connectionist, 19
Correlation, 213
coefficient, 14, 153
degree of, 101, 153, 158
integral, 14, 133, 134, 142, 143, 214
matrix, 164, 165, 191, 218
squared, 153, 154
Correlation dimension, 14, 128, 129, 130, 131, 132, 133, 134, 135, 136, 137, 138, 140, 141, 142, 143, 144, 214
attractor geometry and dynamics, 137
calculating, 132
computational alternatives, 138
factors affecting choice of parameters, 134
Creating independent variables, 150

Cross spectral density, 153, 155
Crossover, 76, 78, 79, 82, 86, 87, 215
Cross-validation, 35, 36, 38, 40, 49, 68, 72, 97, 111

Data
analysis, v, vi, 1, 3, 13, 15, 52, 73, 80, 138, 175, 182, 183, 203, 204
categorical, 44
chaotic, 67, 70, 109, 124, 126
collection, v, 171, 173
compression, 102
conditioning, v, 171, 180, 181, 182, 183, 199, 201
EEG, 55, 56, 58, 59, 74, 141, 142, 177, 181, 182, 205, 207, 212
experimental, 55, 139, 177, 180, 195, 203, 208
features, 22, 41, 180, 181, 203
formats, 176
inspection, v, 171, 177
noisy, 134, 138, 142, 196
numerical, 1, 3, 5, 9
predicted, 101, 102
preprocessing, 2, 3, 56, 177, 180, 182, 203
reduction, 180, 182, 199, 201
specification, v, 171
transformations, 183
Decision
boundary, 17, 18, 39, 172, 182
surface, 17, 122
DeCoster, 136, 140
Derighetti, 137, 139
Desktop options, 10
Destexhe, 139, 140
Determinism, 130, 140
Deterministic, 70, 130, 136, 138, 140
Detrend, v, 183, 184, 185, 203
Ding, 137, 140
Dissipative system, 130
Distributed representation, 51
Dow, 38, 74
Duke, 138, 141
Dvorak, 134, 140

Dynamical system, 2, 14, 129, 130,
      131, 137, 138, 141, 142, 144, 213

EEG, 45, 55, 56, 57, 58, 136, 138, 173,
      181, 184
Eigenvalue, v, 2, 14, 146, 147, 149,
      150, 152, 192, 193, 195, 196, 197,
      200, 201, 214
   estimating system complexity, 147
Eigenvector, 14, 139, 146, 147, 150,
      191, 192, 214
Embedding, 14, 135, 139, 214
   space, 139, 214
Enochson, 154, 159, 160
Entropy, 160, 161, 215
   joint, 161
Epilepsy, 55, 58, 138, 140
Equation
   cubic map, 110, 111, 112, 126, 127
   difference, 67, 100
   differential, 70, 127, 135, 179
   logistic map, 67, 70, 100, 123, 124,
      125, 129, 131
   Lorenz flow, 70, 71, 127, 128
   Rossler flow, 135
Error
   derivative, 29, 30, 31, 33, 34
   surface, 26, 27, 28, 29, 33, 34, 51,
      53, 64, 75, 76
Essex, 132, 136, 140
Eubank, 139, 140
Evaluation function, 76, 79
Evolution, 75, 84, 128, 215
Evolve
   a network, 81
   a population, 13, 83

Farmer, 130, 139, 140
Fast Fourier transform (FFT), 14, 144,
      145, 206
Feedforward, 1, 13, 21, 25, 34, 59, 74,
      83
Filter
   frequency, 187, 194

principal components, 193, 194,
      195, 197, 199
Fitness, 76, 77, 78, 79, 80, 82, 85, 86,
      87, 215
Flotzinger, 58, 73, 74
Ford, 130, 140
Fourier transformation, 55, 56
Fractal dimension, 2, 128, 137
Fraleigh, 153
Frank, 18, 74, 132, 136, 138, 140
Fraser, 135, 140, 161, 163
Fulmer, 135
Funahashi, 25, 41, 73
Function approximation, 13, 24

Gabor, 55, 58, 74
Gaussian, 64, 66, 106, 118, 122
Generalization, 32, 45, 73, 121
Generalized regression neural network,
      1, 13, 91, 93, 94, 95, 215
Genetic algorithm
   approach, 13, 75, 77, 81, 83
   using, 80
Genetic network, 10, 13, 41, 52, 82,
      83, 84, 85, 86, 87, 88, 89, 90, 96,
      117, 118, 119, 121, 122, 123, 124,
      125, 126, 127, 128
Genetic operators, 76, 82
Gibson, 139, 140
Glorieux, 137, 141
Goldberg, 78, 79, 91
Gollub, 70, 73
Grassberger, 132, 133, 136, 137, 138,
      139, 140, 141
Gray, 161, 163
Grebogi, 137, 140, 141
Grozinger, 56, 58, 74

Haussler, 40, 73
Heaviside function, 133
Hebb, 16
Hebbian learning, 16, 19
Hennequin, 137, 141
Hobbs, 130, 141

Holland, 75, 78, 79, 82, 91
Holmes, 126, 179
Honig, 75
Hornik, 25, 41, 74
Horvath, 55, 58, 74

Information
  mutual, 2, 14, 135, 140, 141, 160,
    161, 162, 163, 216
  theory, 160

Jackknifed classification, 43, 97
Jackson, 130, 141
Jando, 55, 58, 74

Kalcher, 58, 73
King, 138, 139
Kloppel, 56, 57, 58, 59, 74
Kohonen, 57, 74, 103, 105, 113

Lag, 14, 132, 134, 135, 139, 166, 167,
    168, 169, 213
Lateral inhibition, 103
Learning
  competitive, 103
  distributed, 19
  function, 116, 121
  machine, 19
  network, 24, 55, 56, 122, 164
  supervised, 59
  unsupervised, 103
Learning rate, 30, 31, 58, 63, 104, 105,
    106, 109
  fixed, 63, 105, 109
  variable, 63, 109
Learning vector quantizer (LVQ), v, 1,
    10, 57, 58, 102, 105, 106, 107,
    110, 111, 119, 121, 122, 123, 216
Lefranc, 137, 141
Leibert, 135, 141
Lemieux, 132, 140
Levin, 80, 91
Linearly separable, 17, 19
Lisboa, 39, 51, 52, 73, 74, 172
Local error, 31

Logistic function, 23, 24, 32, 50, 67,
    100, 124
Lookman, 132, 140
Lorenz, 70, 71, 72, 73, 74, 127, 128,
    135
Lowe, 58, 75
Lyapunov
  exponents, 130, 131, 132, 138
  spectrum, 131

Manglis, 75
Martin, 39, 74
Martinerie, 135, 141
Matrix, 10
  correlation, 164, 165, 191, 218
  covariance, 139, 169, 170, 191, 197
  eigenvalue, 191, 193
  eigenvector, 192
  Jacobian, 52, 53
  scree, 193
  sensitivity, 51, 52, 53, 61
  spectrum, 145, 206
  standardized component scores, 196
McClelland, 19, 74
Mees, 135, 139, 141
Mehridehnavi, 39, 74
Minimum
  global, 28, 29, 82, 215
  local, 24, 28, 29, 78, 79, 82, 213
Minsky, 19, 74
Mitchell, 136, 140
Mohl, 58, 74
Momentum, 30, 31, 34, 64
Moon, 130, 133, 141
Morgan, 75
Muench, 139
Mutation, 76, 77, 79, 80, 82, 87
Mutual information, 2, 14, 135, 140,
    141, 160, 161, 162, 163, 216

Neighborhood function, 106
Nerenberg, 132, 136, 140
Network application examples, v, 116
  learning phase relationships, 122
  learning the XOR function, 120

running, 116

Network functions in Simulnet, 13

Neural network, 1, 13, 21, 25, 34, 55,
  56, 57, 58, 59, 70, 91, 94, 95, 217
  architecture, 21, 42
  as data analytic techniques, 49
  assessing results, 210
  configuration, 32, 53, 81, 107
  creating a training file, 208
  creating training and testing
    matrices, 45
  exemplar, 43, 44, 45, 48, 68, 97,
    100, 111, 113, 172, 182, 208
  exemplar files, 45, 46, 207, 208
  exemplar vector, 20, 44, 94
  function approximation, 24
  hidden level, 25, 26, 32, 33, 39, 40,
    41, 42, 50, 53, 60, 61, 63, 81, 84,
    94
  information displays, 117
  input level, 21, 22, 26, 32, 53, 94
  input node, 21, 22, 23, 25, 32, 34,
    40, 45, 50, 55, 56, 57, 58, 60, 63,
    81, 87, 91, 92, 94
  internal representation, 37, 51, 54,
    75
  learning, 24, 55, 56, 122, 164
  linear output nodes, 31
  logistic output nodes, 31
  models, 16, 19, 20, 79, 107
  number of hidden levels, 25, 39, 40,
    41, 60, 61, 81, 84
  number of hidden nodes, 24, 25, 38,
    39, 40, 41, 42, 60, 61, 91
  number of testing exemplars, 43
  number of training exemplars, 39,
    40, 91, 94, 172
  output level, 21, 24, 26, 32, 33, 34,
    50, 63, 66
  output node, 25, 50
  performance, 38
  restarting training, 119
  structure, 21
  summary of operation, 53
  test phase, 118

test vector, 80, 92, 93, 104, 105, 106
testing data, 60, 67, 68, 71, 72, 97,
  100, 110, 111, 123, 124, 125,
  126, 127, 128
testing error, 36, 62, 68, 69, 72, 84,
  88, 89, 111
testing exemplars, 35, 37, 43, 49, 54,
  56, 57, 60, 68, 69, 70, 72, 87, 88,
  89, 90, 91, 92, 93, 94, 95, 97,
  104, 108, 112, 113, 114, 116,
  119, 120
testing matrix, 42, 45, 49, 60, 72, 94
training, 25
training data, 27, 29, 30, 37, 38, 39,
  45, 51, 59, 63, 68, 87, 88, 96,
  103, 104, 105, 109, 111, 118,
  121, 123, 124, 125, 172, 173,
  181, 204, 216
training error, 36, 62, 68, 69, 72, 84,
  88, 89, 111
training exemplars, 13, 21, 22, 24,
  25, 26, 29, 30, 32, 33, 35, 36, 37,
  38, 39, 40, 42, 43, 45, 49, 54, 57,
  58, 60, 63, 69, 70, 72, 87, 88, 89,
  90, 91, 93, 94, 95, 96, 97, 103,
  104, 108, 112, 116, 119, 171,
  172, 207, 210
training file, 59, 117, 207, 208
training matrix, 42, 45
training phase, 36, 38, 53, 54, 58,
  118, 120, 215, 217
training task, 27, 42
training time, 172
training variables, 172
training vector, 92, 93, 103, 104,
  105, 106, 215, 216, 217
weight vector, 104, 105, 106, 218
weights, 52
Neuron, 15, 16, 17, 18, 21, 40, 55, 64,
  65, 129
Neuronal networks, 16, 19, 20, 21
Nyquist frequency, 174

Osborne, 138, 141
Otnes, 154, 159, 160

Ott, 137, 140, 141
Outliers, 59, 177, 180, 184, 204
Overtraining, 37, 38

P300, 56, 75
Papert, 19, 74
Parallel distributed processing (PDP), 19, 74
Parzen, 102
Pelikan, 137, 141
Peltoranta, 58, 74
Perceptron, 18, 19, 74
Perceptron convergence theorem, 19
Pfurtscheller, 58, 73, 74
Phase
    angle, 15, 88, 90, 122, 159, 217
    difference, 56, 217
    space, 14, 132, 133, 134, 135, 139, 190
Politi, 137, 139
Population
    configuration, 84
    of networks, 81, 82, 83, 87, 88, 89
    parent, 77, 82, 215
    vector, 76
Power
    density, 155
    spectral density, 155
    spectrum, 153, 155, 156
Predicting
    chaotic data, 67, 70, 109
    cubic map, 126
    earthquakes, 48
    logistic map, 123, 124
    Lorenz flow, 127
Prediction task, 122, 123, 124, 127
Principal component
    data reduction, 199, 201
    filtering, 193, 194, 195, 197, 199
Pritchard, 138, 141
Probabilistic network, v, 10, 13, 43, 91, 92, 96, 97, 98, 100, 101, 117, 118, 120, 121, 122, 123, 124, 125, 126, 127, 128, 202, 209, 210

comparing with the backpropagation network, 94
Probability density function, 93, 94, 95, 97, 102
Procaccia, 132, 133, 136, 138, 139, 140, 141
Provenzale, 138, 141
Pulse
    repetition rate, 17
    train, 17

QL algorithm, 191

Ramsay, 75
Rapid-eye-movement (REM), 56, 74
Rapp, 135, 139, 141
Ravani, 137, 139
Recurrent network, 21
Rigler, 75
Romeiras, 137, 141
Roschke, 56, 74
Rosenblatt, 18, 19, 74
Rothman, 136, 141
Rubio, 137, 139
Ruelle, 136, 140
Rumelhart, 74

Sayers, 136, 141
Schaffer, 135
Schema, 78, 79
    theorem, 79
Schuster, 135, 141
Schwartz, 139
Search space, 75, 76, 78, 79, 81, 82, 83, 119
Self-organizing map (SOM), 10, 103, 104, 105, 107, 216, 217, 218
Sensitivity matrix, 51, 52, 53, 61
Sepulchre, 140
Session log, 8
Seyal, 55, 58, 74
Shannon, 160, 163, 215
Siegel, 55, 58, 74
Sietsma, 38, 74

Sigmoidal function, 18, 216, 217
Silva, 75
Simulnet exercise
  computing a Fourier spectrum, 144
  computing coherence and phase, 156
  computing correlation dimension, 141
  computing eigenvalues, 151
  computing mutual information, 161
  data classification, 98
  detrending a matrix, 185
  Fourier analyzing the data, 206
  inspecting the data, 205
  learning a functional relationship, 87
  neural network classification, 207
  opening a file as a matrix, 11
  predicting chaotic data, 67, 70, 109
  predicting chaotic data 1, 67
  predicting chaotic data 2, 70
  principal component data reduction, 199, 201
  principal component filtering, 194, 197
  standardizing a matrix, 186
Singular value decomposition, 138, 139
Siska, 134, 140
Slater, 58, 75
Smith, 56, 136, 141
Smoothing, 94, 95, 98, 99, 101, 102, 111
Specht, 91, 102
Spectral power, 57
Spectrum
  frequency, 14, 206
  Lyapunov, 131
  power, 138, 144, 145, 146
Standard
  deviation, 59, 60, 66, 95, 184, 185, 186, 204
  error, 142, 143, 154, 155
Standardize, 185, 186, 204
Starting Simulnet, 5
Stinchcombe, 25, 74

Stochastic, 51, 136, 138, 141, 171, 190, 193, 194
Substantive question, 15
Supervised
  classification algorithm, 216
Swinney, 135, 140, 161, 163
Synapse, 16, 18, 21, 26

Takens, 132, 134, 135, 141
Thornton, 41, 75
Threshold function, 18
Toolbar buttons, 7, 8
Training
  supervised, 104
  unsupervised, 105
Truty, 135

Universal function approximator, 1, 25, 50, 81, 96, 106

van Dijk, 51, 73
Variable
  criterion, 37, 42, 44, 45, 46, 49, 62, 85, 91, 94, 118, 124, 125, 127, 208
  predictor, 37, 39, 42, 44, 45, 48, 91, 94, 118, 124, 125, 127, 182, 201, 202
Variable transformations, 137
Variance
  noise, 196
  signal, 196
Vector
  codebook, 102
  criterion, 44, 54
  population, 80, 81
  predictor, 44, 53, 54
  quantization, 103
  reference, 103, 104, 105, 216
Verbeke, 51, 73
Vogl, 75

Wasserman, 91, 94, 102
Webb, 58, 75

Weight space, 27, 28, 29, 30, 75, 213
Werbos, 19, 75
White, 25, 74
Wu, 75

XOR function, 117, 120, 121, 122

Yilmaz, 130, 139, 140
Yorke, 137, 140, 141

Zink, 75